teach yourself®

owning a horse

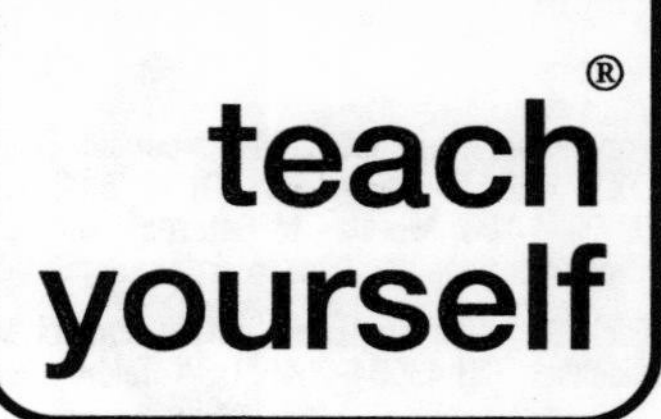

owning a horse

carolyn henderson

For over 60 years, more than 50 million people have learnt over 750 subjects the **teach yourself** way, with impressive results.

be where you want to be
with **teach yourself**

For UK order enquiries: please contact Bookpoint Ltd, 130 Milton Park, Abingdon, Oxon, OX14 4SB. Telephone: +44 (0) 1235 827720. Fax: +44 (0) 1235 400454. Lines are open 09.00–17.00, Monday to Saturday, with a 24-hour message-answering service. Details about our titles and how to order are available at www.teachyourself.co.uk

For USA order enquiries: please contact McGraw-Hill Customer Services, PO Box 545, Blacklick, OH 43004-0545, USA. Telephone: 1-800-722-4726. Fax: 1-614-755-5645.

For Canada order enquiries: please contact McGraw-Hill Ryerson Ltd, 300 Water St, Whitby, Ontario, L1N 9B6, Canada. Telephone: 905 430 5000. Fax: 905 430 5020.

Long renowned as the authoritative source for self-guided learning – with more than 50 million copies sold worldwide – the **teach yourself** series includes over 500 titles in the fields of languages, crafts, hobbies, business, computing and education.

British Library Cataloguing in Publication Data: a catalogue record for this title is available from the British Library.

Library of Congress Catalog Card Number: on file.

First published in UK 2006 by Hodder Education, 338 Euston Road, London, NW1 3BH.

First published in US 2006 by The McGraw-Hill Companies, Inc.

This edition published 2006.

Typeset by Transet Limited, Coventry, England.
Printed in Great Britain for Hodder Education, a division of Hodder Headline, 338 Euston Road, London, NW1 3BH, by Cox & Wyman Ltd, Reading, Berkshire.

Impression number 10 9 8 7 6 5 4 3 2 1
Year 2010 2009 2008 2007 2006

contents

introduction

Many people dream of owning a horse or pony. For some, particularly those who are born into equestrian-minded families, it starts and may even come true in childhood. For others, the chance to learn how to ride – and from that, the idea of a horse of their own – comes later in life.

The world of horses and riding has, thankfully, cast off most of its elitist image. These days, you no longer need to be in jodhpurs before you are out of nappies to make riding part of your life. Nor do you need to own acres of land to be able to have a horse of your own; many of today's owners live in towns and pay to keep their horses in rented accommodation on a livery yard, either travelling to care for them each day or paying the yard owner to do all or part of the work.

Whilst owning any animal is a responsibility, whether it be a horse or a hamster, owning a horse is particularly demanding. It takes time, dedication, skill and money and becomes not so much a hobby as a way of life. This book will not only help you gain an understanding of the practical skills, it will help you decide whether or not becoming a horse owner will fit into your lifestyle and other commitments. (Though for many people, owning a horse becomes a lifestyle in itself!)

Although it contains information about all the areas potential horse owners need to consider, no book can be a substitute for hands-on experience gained under knowledgeable supervision. There are those who buy a horse – or, more usually, a pony for their children – with no idea of their needs or natures, but this is a mistake. It can and often does have serious consequences and, inevitably, the one who suffers is the horse. The International League for the Protection of Horses, a world leader in rescuing and rehabilitating horses, says that the biggest enemy of equine welfare is not deliberate cruelty, but ignorance.

If you already have experience of riding and helping to look after horses, this book will act as a back-up and help you widen your horizons; if it is some time since you were directly involved with them, it will bring you up to date with everything from modern methods to scientific advances. That does not mean that 'new' always means 'better', more that we now know more about why some approaches work and some don't.

For those with little or no experience, but who have always wanted to gain more than a general admiration of a beautiful and fascinating animal, it will explain the nature of horses and how to deal with and look after them responsibly, fairly and safely. At the end of it, you may even decide to go for it and enter the world of horses, with the ultimate aim of owning your own.

It will also help parents or guardians of children eager to become pony owners. No matter how dedicated and competent the child, adult help is essential for practical and safety reasons and, of course, to finance the pony and its care.

Throughout the book, a horse will be described as he, though this is simply for ease of description and means no disrespect to the female of the species. The alternative 'it' does a disservice to the extent to which most owners care about and for their horses.

Horses and ponies

Although this book is called *Teach Yourself Owning A Horse*, it applies equally to anyone wanting to learn about owning a pony. The basic difference between the two is height as, in general terms, a pony is a small horse.

Both are measured by dropping a perpendicular line from the withers (at the base of the neck) to the ground. Traditionally, measurement is made in hands and inches. A hand is 4 inches (10 cm) so a pony of 12 hands 2 ins high, expressed as 12.2 hh would measure 4 ft 2 ins and a horse of 16.2 hh would measure 5ft 6ins. This system is still widely used, though metric measurements are also employed, particularly in competition – for instance, classes may be open to ponies not exceeding 148 cm, the equivalent of 14.2 hh. Just to make it even more interesting, the Shetland pony, which is the smallest of the breeds native to the UK, is traditionally measured in inches!

As a general guideline, a pony measures up to and including 14.2 hh and over that, is classed as a horse. Again, there are the odd exceptions. Two ancient breeds, Arabians and Icelandics, are always called horses no matter what their height and some competitive disciplines class animals up to 15 hh as ponies.

Getting the basics

A horse is usually a working animal that is ridden, driven or both. Although he will probably become an important part of your life, he will be more than a pet. Even if you ride purely for pleasure and never compete, he is an athlete who has to be healthy and fit to carry out his job. Competitive riders can, of course, get equal pleasure from their riding, but it is important to realize that the horse whose work consists of an hour's leisurely hack most days deserves the same care and attention to detail as the one taking part in demanding sports.

Some horses and ponies are kept in retirement or, perhaps because injury means they can no longer be worked, as companions to others. Even if you don't want to ride and are perhaps considering keeping a retired pony, he will have similar needs – or rather, he and his companion will, as it is unfair to keep herd animals such as horses on their own.

If you are a rider, or buying a pony for your children, you and they should be confident in handling horses in all everyday situations. Even if a child takes the main active role, there may be times when a parent needs to take over to cover, because of illness or other commitments.

Riding skills should be at the stage where you are at least confident and competent in walk, trot and canter on a well-schooled horse, both alone and in company, and are hacking out (called trail riding in the USA) with equal competence and confidence and a knowledge of road safety. This is the bare minimum of experience required before taking on a horse, and whilst all owners should be prepared to continue their training – because with horses, you never stop learning – at this stage it is essential.

You should also understand the principles of feeding and management, how to tell when a horse may have a health problem, basic first aid, and the use and fitting of tack and equipment. Other areas which need to be studied range from the care of the horse when being transported, to how to assess different types of animal so you know what sort to look for, under expert guidance.

The practical parts

All the above will help enormously if you or someone else in your family want to ride regularly. If you want to own a horse, you also need to be sure that you can cope with the demands this will make on your time and bank balance.

Looking after a horse yourself is the cheapest way to do things and also the most rewarding, because the more time you spend with him, the better you will get to know him. However, it will take up a lot of time; you will need to see him once or twice a day, depending on how he is kept, and factor in time for routine care as well as exercising. Expect to spend at least one or two hours a day with your horse, excluding travelling time.

Paying someone else to look after your horse all or part of the time and, if necessary, exercise him when you are unable to will not give you the same satisfaction as caring for him yourself. However, for some people with demanding jobs this is the only way to have a horse of their own and ensure that his needs are always met. It is the most expensive way of owning a horse.

Whatever system you intend to follow, regular costs, not including the initial purchase price, will include:

- rental of stable and grazing
- feed and bedding
- shoeing every six weeks
- tack, equipment and rugs – purchase and maintenance
- insurance for third-party liability and veterinary fees (recommended)
- annual vaccinations
- dental care

Costs vary, but as an example, expect to pay £15–25 to rent a stable and grazing at a livery yard; £100–160 per week for someone to look after your horse and a pro rata sum if you do part of the work; £45–65 every six weeks for shoeing. Although it is, not surprisingly, cheaper in some ways to keep a small pony than a competition horse, costs such as veterinary care and shoeing will be the same.

It will be immediately obvious that a family pony will cost a lot more to keep than a family dog. However, most horse owners are prepared to go without other things to pay for the satisfaction and enjoyment they get from owning a horse.

Getting a feel for it

The best approach for any would-be horse owner is to get practical experience of looking after a horse before buying one. Not only will this give you confidence, it will show you whether it is a realistic ambition or whether you need to wait until your circumstances or finances change.

Many riding schools offer horse management courses, giving you the chance to learn practical skills under qualified supervision. These can provide an excellent grounding; if you choose a school that is registered with the Association of British Riding Schools or the British Horse Society, you know that its teaching will follow safe and generally accepted practices. Some schools also offer leasing schemes, where you pay a fee to have the use of a suitable horse for part of the week; he remains the property and responsibility of the school and there will be restrictions on what you can do with him, but you should also be able to get involved in routine care.

Colleges who offer equine courses often include part-time and evening courses for would-be owners. Recently, the International League for the Protection of Horses introduced lectures on responsible horse ownership at its centres in England and Scotland, highlighting how much it costs to keep a horse and the time commitments that it demands as well as giving information on issues such as suitability of horse and rider, daily routine, feeding, watering, pasture management, stabling, and professional support such as farriery and veterinary care.

An obvious way of getting practical experience is to offer to help someone who owns a horse. Many local publications carry adverts offering riding in return for help with work and, perhaps, costs, but these arrangements should be approached with caution. They can be risky unless you are sure that the owner is knowledgeable and the horse is suitable for a rider of your standard – you may end up being asked to do things incorrectly and/or find yourself riding a horse that is too much for you.

Whatever avenue you follow to gain hands-on experience, it is a good idea to check that the owner of the horse has adequate insurance cover, in particular for third-party liability.

This should automatically be the case with approved riding schools and colleges but not all private owners are aware of its importance. You may also want to take out personal insurance against accident and injury; some specialist equestrian insurance companies offer policies tailored especially for riders who do not own their own horses.

the nature of horses

In this chapter you will learn:

- **how horses react**
- **what is equine body language**
- **how to influence horses with your body language**
- **what are basic handling skills.**

It has been said that horses bite at one end, kick at the other and that what happens in the middle is equally unspeakable! Fortunately, that isn't true: horses are rarely aggressive towards people – and those that are have usually been made that way through bad handling, bad experiences, underlying pain or a mixture of these causes.

However, horses – like dogs – are domesticated wild animals and even the cutest pony will retain his wild horse instincts. The more you understand about how horses think and behave, the better and safer you will be around them.

Fossils show that the ancestors of modern horses and ponies first appeared about 60 million years ago. The first horse, Eohippus, was about 36 cm high and had four toes on each front foot and three on each hind foot. He fed on leaves, since there was no grass at that time. As the Earth's landscape changed from swamp to plains, so the horse developed longer limbs and his feet changed into hooves. By the time horses were first domesticated, about 6,000 years ago, there were four distinct types of horse who gradually developed into distinct breeds. Today there are more than 75 breeds of horses and ponies throughout the world.

Senses and sensibility

When it comes down to basics, the horse is by nature a prey animal just as the human is a natural hunter. This means that when a horse feels frightened or threatened, his natural reaction is not to attack, but to run away. His speed is his best defence and even the slowest, lumbering horse can move surprisingly quickly if he has to.

As a prey animal, his eyes are sited at the side of his head rather than at the front (as is the case with the hunter). He has a much wider field of vision than we do and only has to turn his head slightly to get an all-round view, though he also has 'blind spots' directly behind him and directly below his nose, and has to raise his head and point his nose forward to see things in the distance.

His upright, mobile ears channel an acute sense of hearing. We know that horses can hear frequencies much higher or lower than those which people can pick up and can also hear across much greater distances. His sense of smell is also much keener than ours and a horse who raises his head and gives a loud, sharp snort is taking in messages through smell.

A horse is sensitive to touch and, in particular, uses his muzzle and lips to investigate things he can't see because they are in the blind spot below his nose. This is how a grazing horse picks out which grasses and plants he wants to eat. If you watch horses in the field, you will also see how they enjoy mutual grooming, moving their teeth along each other's necks and backs.

Horses are by nature herd animals rather than solitary. In every group, there is a pecking order and every time a new horse is introduced, he will have to find his place on the ladder. When mares and geldings live in mixed groups, a mare will usually be the boss. Even in the wild, when a herd is led theoretically by a stallion, a dominant mare will have as much if not more control over other herd members.

Although all horses have the same instincts, they are also individuals with different temperaments. Genetic predisposition, age and experience will often result in common denominators, but there are always horses that break the mould!

In general, a horse with a high percentage of Thoroughbred or Arabian blood will be more sensitive, both physically and mentally, and quicker to react than one whose parentage descends from the heavier horse breeds. Young horses, like all young animals, are more likely to be playful, especially with their equine companions. A horse who has been introduced considerately to varying sights and sounds will be more likely to take new ones in his stride; positive experience brings tolerance.

Geldings (castrated male horses) are generally accepted to be easier to deal with than mares and stallions, who are both 'entire' horses with nothing taken away. Again, there are always exceptions! Stallions are not wild animals and should not be aggressive, but are not suitable for the average owner. Mares can be more temperamental when in season but can also be very rewarding for the right people. The old saying that you tell a gelding, ask a mare and discuss it with a stallion has a lot of truth in it.

Handling by instinct

So how does understanding why a quiet, well-behaved horse is a wild animal at heart help you become a better person around horses? There is nothing mystical about it and certainly it has nothing to do with 'horse whispering' and other mysterious arts, which arguably exist only in the imagination.

It does make you more aware of how the way you talk, move and behave around horses influences their reaction to you and how they react to their environment. For instance, horses dislike loud noises, so don't shout; nor is it a good idea to play loud radio music on a yard, though many do. However, horses do respond to the voice when it is used in a soothing or encouraging tone, as appropriate – just don't chatter away to your horse all the time or he will simply switch off.

Sudden movements will startle him and his reaction will be to take flight – which, depending on the horse and the circumstances, may mean anything from pulling back and breaking his headcollar rope to galloping off across the field.

Unfamiliar or worrying smells may make him nervous. For instance, many horses are frightened by the smell of pigs and some are worried by the smell of oil or disinfectant on clothing.

If you approach a horse directly from the front or directly behind, he won't see you clearly until you are very close to him, so may be startled. Instead, approach him from the side whenever possible.

It is a good idea to speak quietly to a horse before you touch him, especially if there is a risk that you have approached from a blind spot and he is not aware of you, or he is dozing. Touch him with confident, sweeping movements: don't grab or dab at him.

Some horses like being stroked and, in particular, being scratched at the base of the neck and the withers, a favourite spot for two horses indulging in mutual grooming. They may also enjoy having their forehead rubbed.

However, most do not particularly like being patted, though they may tolerate it, and they certainly do not like being patted on the head. Some horses enjoy being fussed whilst others prefer to keep their distance, or will show you when they appreciate contact. Frustrated 'touchy feely' owners with horses who prefer to keep themselves to themselves should put their enthusiasm into grooming, instead (see Chapter 05).

Body language

Horses communicate with each other through body language. By understanding their signals, you can assess how a horse is feeling – alert, dozy, bad tempered and so on – and also modify your body language to influence him.

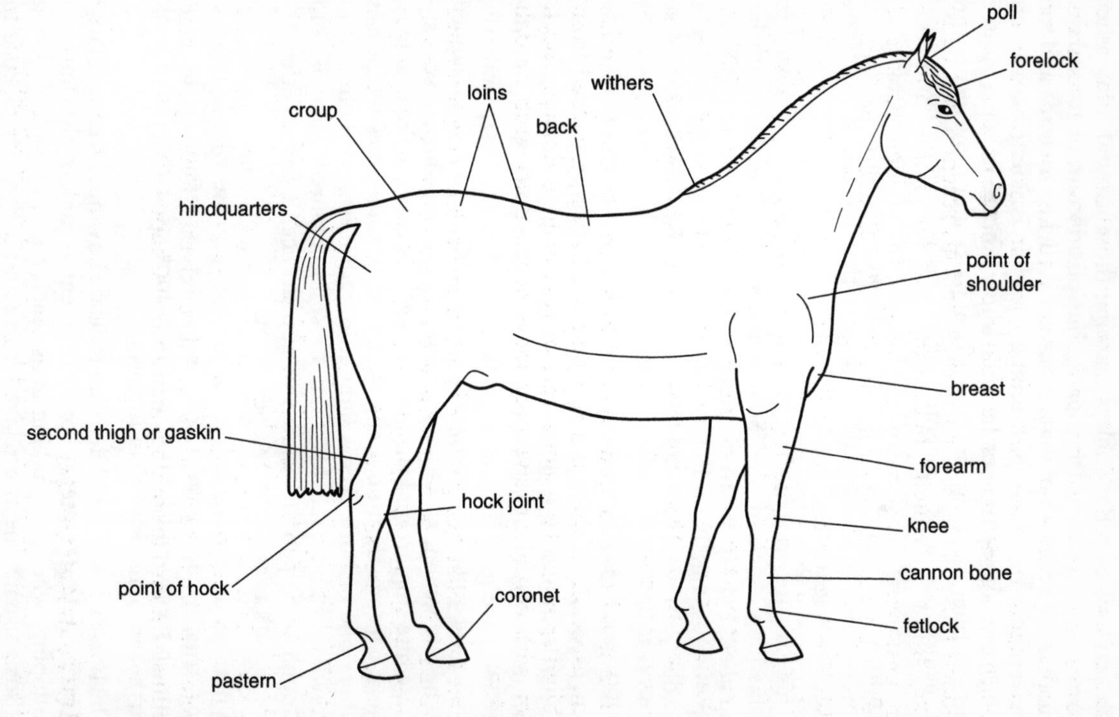

figure 1 points of the horse

Observant horse owners read their horse's whole body, but the clearest signals come from the head. The ears, eyes and lips all tell you how the horse is feeling. Most people know that a horse who lays his ears flat back is warning those around him, either horses or humans, to stay back, but positioning of the ears is much more subtle than that. If one ear is flicked forward and the other back, the horse is listening to what is going on around him; you can sometimes see this when a horse is being ridden and his rider is asking him to perform particular movements and it means that he is concentrating on his rider. The horse who goes round the school with his ears pricked, looking as if he is ignoring any signals, probably is doing just that, unless he is approaching a fence and looking at the obstacle he is being asked to negotiate.

If his ears are pricked sharply forward, the horse is interested in something ahead of him. Sometimes, you will see him with his ears held loosely out to the side; if he is fairly alert, he is picking up sounds, but if he is relaxed in his body, perhaps resting a hindleg and with his head down, he is resting. However, a horse who is feeling ill will often show similar body language, so it is important to be able to recognize the difference (see Chapter 09.)

It may often be said that a particular horse has a 'kind eye'. This is partly due to natural conformation, as a large eye is more attractive than a smaller one even though the horse in the second case may be just as good natured as the first. Similarly, whilst the horse who shows the white of his eye is often assumed to be frightened, some horses have a more pronounced sclera – the white ring around the eye – which will often make him look more nervous than he really is. You may sometimes see horses where one or both eyes is light coloured; it is more common in skewbalds and piebalds than in solid-coloured animals and though some people dislike this appearance, it has no effect on the horse's eyesight or temperament.

Whatever the size or colour of the eye, a horse who is relaxed and friendly will have a soft, inquiring expression. If he is in pain, his eyes will be dull and may look as if they have sunk farther into his head than is normal.

The horse's nostrils and lips also indicate how he is feeling. Flared nostrils indicate interest or excitement – unless the horse is galloping, when he is taking in as much air as possible. Relaxed nostrils usually mean a relaxed horse and if they are pinched in and wrinkled he is either showing distaste, feeling ill or indicating aggression; you need to put together the overall picture, sometimes very quickly. A dozing, totally relaxed horse will have loose lips, sometimes to the extent that the bottom one is hanging down. Tight lips are often a sign of a tense or nervous horse.

Watch the way a horse holds his tail, too. Although some breeds, notably the Arabian, have a naturally higher tail carriage than others, a horse who is excited and generally feeling happy with life – perhaps because he has just been turned out in the field with his friends – will often lift his tail and generally show off, taking high, exaggerated steps to say 'Look at me'. But if he is frightened or ill, his tail will more likely be clamped down.

A horse who is relaxing quietly in his field or stable will often rest a hindleg and let his head and neck hang down; again, this may also be the body language of one who is ill. A frightened horse may lash out with a hindleg if frightened by sudden movement or noise from behind, but one who lifts a hindleg is giving a warning: watch out or I'll kick. Impatient horses often paw the ground with a front foot, perhaps because they know their food is coming or they are reluctant to stand still.

The best way to learn the signals horses offer is to watch groups of them in the field. If one tries to rise above his designated place, he will be put back in it via signals from a superior companion such as pinned back ears, a stretched out neck and opened mouth, meaning: 'Back off or I'll bite.' A horse who is interested in your approach, perhaps because he associates it with something pleasant such as a titbit reward for being caught, will stand or walk towards you with his head lowered and his ears held gently forward.

Consideration when you are grooming and tacking up a horse will usually mean that he accepts what you are doing; many horses enjoy being groomed. But if you pull up a girth too quickly, or use a brush that is too harsh for a thin-skinned horse, he will tell you. Pinned back ears, a head that swings round with the threat or intention to bite or a hindleg raised to kick forward toward the belly is a clear warning to stop it, or else. A horse's reaction to routine procedures may also tell you how he has been treated in the past; they have long memories and if a previous owner or handler has girthed up roughly, a horse may continue to show signs of anticipated discomfort.

Signalling back

Over the past few years there has been increased awareness of how we can influence horses' behaviour by our own body language. Good horsemen and women have always used this skill, even if they have acquired it through instinct rather than instruction, but high-profile trainers – many of whom have equally high-profile marketing operations – have made more horse owners and riders aware of its importance.

However, whilst it might be tempting to look on some individuals as equestrian gurus, it is important to base all your dealings with horses on fairness and safety rather than on mysterious images of 'horse whisperers' who seem to perform miracles. The fact that a gifted trainer with many years' experience of assessing and dealing with horses can get an untrained horse to accept a saddle, bridle and rider within half an hour or even less in front of a large audience does not mean that this is the best way to start a youngster's education, nor does it mean that the horse will automatically understand a rider's signals and be safe to ride outside an enclosed pen.

What it does show is that we can't expect horses to know automatically what we want them to do – though they often make a better job of it than we deserve – and that it is up to us to influence them in ways they find easiest to understand. Behaving fairly and consistently is vital; if you accept the way a horse behaves one day and punish him for doing exactly the same the next because you are not in such a good mood, you are being unfair and inconsistent.

Even the smallest pony is much stronger than even the strongest person, but can be handled and trained so that he does not use that strength against us. Tiny ponies are sometimes much more badly behaved than large horses, simply because their owners have not realized they need to be educated and handled in just the same way.

Horse sense

The best way to communicate with a horse on the ground is through body language that echoes his own. If you stand tall, square your shoulders, face up to a horse and look him in the eye, you are telling him that you do not want him to walk forward into your space; many horses will step back if you adopt this posture. But if you drop your shoulders, turn so that you are half facing away from him and do not look directly at him, you are signalling that you are being non-dominant and he will be more likely to come towards you – using this technique to catch a horse in the field will be much more effective than marching up to him with a headcollar!

Horses react very quickly to body language, which can be used to great advantage as long as you are consistent. You can teach a horse to lead obediently and to stop when you do by using pressure and release – a firm pull on the rope when you stop followed by instant release of the pressure as he obeys will soon have most horses responding to the slightest signal. This kind of positive reinforcement works only if everyone who deals with the horse follows the same system, as if he gets conflicting signals he can't be expected to understand what you want.

A horse will learn through positive reinforcement, not negative. If he does what you ask, a quiet word or scratch on the neck will reinforce the behaviour, but if you shout at him or hit him when he shies at a bird flying out of the hedge, you are reinforcing that there is something to be frightened about. Obviously there will be times when you have to correct a horse, but it must be done immediately, so he knows that the correction is connected to his preceding action, and it must be done calmly. Corrections can be extremely subtle – for instance, you are making a correction when you repeat an instruction that he has ignored or not understood.

In recent years, systems of handling and riding horses that have been marketed in America have found popularity in the UK and Europe. They include Parelli, named after its founder, Pat Parelli; the teaching of Monty Roberts, overseen in the UK by Kelly Marks under the name Intelligent Horsemanship; Think Equus, with UK trainer Michael Peace; the Tellington-Jones Equine Awareness Method; and clicker training, a system that has become widely accepted in the field of dog training. All may pay investigation and have various things to offer, but it would perhaps be just as narrow minded to say that there is one method that must be followed as it would to condemn any of these systems as gimmicky.

Different philosophies are often referred to under the umbrella term of natural horsemanship, which, if taken to mean understanding how horses behave and tailoring the way we handle, care for them and ride them with that in mind, is something good horse keepers have practised for centuries.

One system that can be guaranteed to work is that of the intelligent, fair, consistent owner, no matter what label you want to put on it.

Safe as horses

No matter how well you know a horse, or how quiet he seems, try not to get over-confident or slapdash. The day you walk too close behind him is the day he will be startled by someone dropping a bucket and shoot backwards or kick out.

Similarly, always wear appropriate clothes when handling horses – in particular, gloves and suitable footwear. If a lead rope pulls through your hands when a horse shies, it hurts. It hurts even more if your horse spooks and treads on your toe when you are wearing soft shoes, so wear strong boots. Ones with reinforced toes offer even more protection. There will also be times when it is sensible to wear a hard hat when leading or handling a horse, perhaps one who is excited or a young horse at his first show.

First catch your horse...

Everything we know about the nature of horses can be put into practice when carrying out everyday handling: first catch your horse, then lead him to the yard, then tie him up. Most horses are fairly easy to catch, as long as you approach them in the right way. This is one of the few situations where it is a good idea to reward a horse with a titbit – usually it is not a good idea to feed these, as it can encourage them to become pushy and perhaps to nip.

To catch a horse, approach him from the side at an angle to his shoulder, with the headcollar and rope in one hand; don't swing your arms or the headcollar and keep your shoulders low and your gaze lowered, in the inviting stance explained above. Talk to him as you approach – it is the tone of voice that matters, not what you say, so something as uninspired as 'Good boy' is fine – and when you reach him, run your hand down his neck or shoulder.

Place the lead rope round his neck and gently raise the noseband of the headcollar over his muzzle. Pass the headpiece over his head and fasten it at the appropriate level, then reward him with a scratch, a titbit or both. Be careful about giving a horse titbits if he is part of a group of animals, as they may crowd round if they think food is available.

If a horse has a reputation for being difficult to catch, it is a good idea to make time to visit him regularly in the field without catching him and bringing him in to work. Go up to him as above, give him a titbit if possible, then walk away. When you can do this easily, put a headcollar on, lead him a few strides then take the headcollar off and walk away. Do this every now and then so he does not automatically associate being caught with work. If the horse is in a group and you feel giving titbits is risky, make the effort to take him in or outside the field into safe surroundings and give him his reward, then turn him out again.

Leading and turning out

Traditionally, horses are led from the left-hand side (near side) – in fact, everything from tacking up to putting on and removing rugs to mounting is usually done from this side. No one really knows why this is so and the only explanation is that mounted officers wore their swords on their right-hand side, so needed them to be on the side farthest away from their horses! However, it is far better to accustom a horse to being handled and led from both sides equally, as this makes life easier and means that he is more likely to use muscles evenly on both sides of his body.

When leading him in a headcollar, hold the rope about 15 cm from the clip in the hand nearest to him and the loose end in your other hand. Never wrap the rope round your hand, as if he takes off you could be dragged and injured. Look in the direction you want to go in and walk forward; an educated horse will walk with you without pulling or lagging behind and you should be able to keep a position just in front of his shoulder.

When you turn, turn him away from you. If you turn him towards you, he is more likely to tread on you by accident. When you lead him through a gate, make sure the opening is wide enough so as not to bang his hips. If you are taking him into a field where other horses are present, you will have to keep hold of the gate with one hand and the headcollar rope with the other; in this situation you obviously have to turn his head towards you so that you can close the gate, so position your feet so they don't get trodden on.

Close the gate behind you and turn the horse to face it with his body straight behind. Quietly unfasten the headcollar and step back, watching him for a few moments so you can make sure you are not in range if he decides to whip round, kick up his heels and charge up the field.

If you lead your horse through the gate and find his friends immediately come cantering towards him, it is safer not to turn your back on them but to move a short way from the gate, take off his headcollar and step out of the way. Doing it by the book – this one included – sometimes has to be compromised for the sake of self-preservation, but you should still try to avoid sudden movements which may add to the excitement.

With the horse who is habitually difficult to catch, you may be advised to leave a headcollar on him in the field until you can hopefully make him more amenable, as detailed above. Whilst this makes it easier to give him a reward and clip on a lead rope without having to pass a headcollar over his nose, it should be a last resort – not only does it make it easier for a thief to catch him, but there is also the risk that he will get caught up on something, which can happen in the safest field. If it is unavoidable, use a leather headcollar that will break under stress or one designed for field use with a breakaway section incorporated.

Tying up

All horses should allow themselves to be tied up so that they can be groomed, tacked up and so on. They should be securely tied, but there should also be a breakaway point in case of emergency. Never tie a horse directly to a gate or fence, because if he pulls back in a panic, he could find himself dragging a gate or a fence rail, which would be even more frightening and potentially dangerous.

To be safe, always use a quick release safety knot and tie the lead rope to a loop of strong string or breakable baler twine, attached to the tying-up ring or rail as a deliberate weaker link between the horse and the securing point.

To make a quick release knot:

1 pass the rope through the string loop
2 make a loop in the rope, close to the string
3 put the rope loop across the long part of the rope
4 pull the loose end through and, holding it near the loop, pull the rope taut
5 pass the loose end through the loop.

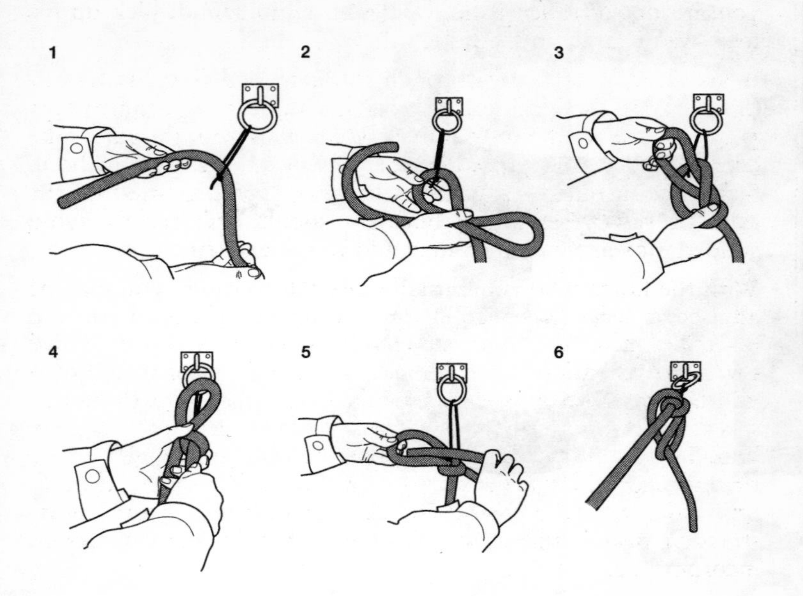

figure 2 how to tie a quick release knot

Some horses, especially those who have not been taught to tie up as youngsters, will deliberately pull back and break the loop. Don't give in to the temptation to tie them directly to a ring; instead, use a stretchy tail bandage instead of string or twine to make the safety loop. This way, the horse doesn't meet a definite resistance when he pulls back and will usually learn to accept being tied up.

02 a home for your horse

In this chapter you will learn:

- **why keeping a horse at livery is usually best**
- **how different forms of livery work**
- **how to find the right yard.**

Before you can think of buying or loaning a horse, you need to find somewhere for him to live. If you are lucky enough to have suitable grazing and buildings of your own, it will be a case of making sure they are in good repair and provide a safe and healthy environment, which is dealt with in the next chapter. But whilst many horse owners would love to be able to keep their horses at home, most have to rely on renting accommodation from other people.

For the first-time owner, this is the best way, as long as you pick the right yard and the right owners. An experienced and knowledgeable yard owner will make sure that overall management is correct and will advise you if there is anything you are unsure or worried about.

In the wild, horses cover up to 25 km each day searching for food. Although the domesticated horse obviously does not have to do this, he needs plenty of time outdoors to keep him happy and relaxed. Although some horses are still kept stabled all or most of the time, this is not a natural lifestyle and most people consider it to be an unfair regime. For most horses and owners, a good system is to turn him out during the daytime and stable him at night.

As more farmers diversify, there is an increasing number offering accommodation for horses. Unless this is owned or managed by someone with specific knowledge of horses, it is not usually the best option for a novice horse owner. Horses have very different needs from animals such as cattle or sheep and problems can result if this is not recognized.

Livery systems

Yards that specialize in offering accommodation for horses, either on a DIY system or with some help available, are called livery yards. Some riding schools also take liveries. There are five main systems, all with pros and cons.

DIY livery means that you rent a stable and grazing and do all the work involved in looking after the horse. It gives the most satisfaction in that you get to know your horse better through having so much contact with him and is the cheapest in initial outlay, though remember to factor in the costs of travelling – and time, if in your case time means money.

Part DIY or assisted livery is a good alternative for owners whose work and/or family commitments mean that they cannot visit their horse twice a day. It means that you agree a division of tasks with the yard owner and accordingly pay a higher livery fee; for instance, you might take care of feeding, mucking out, grooming and exercising in the morning, then turn your horse out and ask the yard staff to bring him in, check him and adjust rugs if necessary and feed him at night.

The alternative is to join forces with another owner on the yard and share the work of looking after two horses, allocating tasks according to your daily schedules. This can work well and saves money, but only if both people are totally trustworthy and can deal with each other's horses happily. In general, it is an option which is better for more experienced owners – when you are starting out, it is better to concentrate on your own horse and have the security of knowing that someone on site will look after his needs when you can't.

If you are considering either of these systems, you must have emergency back-up, either from the yard, another owner or a family member. Persuading a non-horsey member of the family to learn how to do the basics will pay dividends if, for instance, you are ill – and if you're lucky, he or she will also become interested!

Full livery, where the yard staff are totally responsible for your horse's care and well-being, is the opposite end of the scale from DIY. Some yards will include absolutely everything when necessary, whilst others will expect you to exercise your horse and be responsible for tack cleaning. It is the most expensive way of keeping a horse and even those who find it the only viable option will usually want to spend time with their horses and groom them when time allows, perhaps at weekends.

Working livery is a system offered in some riding schools and basically means that you share both the use of your horse and the cost of his keep. It may mean that he is used for staff or school lessons on certain days of the week whilst you ride him on the others, in return for reduced costs and often help with looking after him.

Although there can be advantages with a carefully worked-out system – for instance, some schools may only allow riders who are as experienced or more experienced than you to ride your horse – there are, in general, more potential disadvantages. You will perhaps feel that your horse does not belong to you properly and there may also be occasions when you want to ride him but are unable to do so.

Grass livery is a term applied when a horse or, more usually, a pony, is kept in a field all the time. Many animals are perfectly happy living out 24/7 as long as the grazing is correctly managed and they have shelter available in all weather conditions, but it is essential that a stable should be available in cases of illness or injury. A good yard owner will appreciate this.

This will usually be the cheapest form of livery, but will involve just as much work. It may also influence the type of horse you buy, because though some horses with fine coats – notably Thoroughbreds and those with a high percentage of Thoroughbred blood – can live out successfully in well-managed conditions, not every yard can provide them. Native ponies and horses with a high percentage of native or draught blood, such as cobs, are usually hardier.

The right yard

Finding the right yard is rather like finding the right home for yourself. Not only do you need to make sure that your horse will be safe and comfortable, it is important that you will be happy there, too.

For instance, if you are an adult rider wanting some peace and quiet with your horse, you might not want to keep him on a yard that specializes in children's ponies. Although most children who love their ponies are dedicated to looking after them, some adult owners are happier in the company of other adults. Similarly, if you are looking for a home for a child's pony, you would probably feel out of place on a yard catering for competition horses, no matter how nice the people. However, most yards cater for a mix of owners and a good yard manager will ensure that everyone adheres to the same rules.

If you have learned to ride at a riding school, you may be able to obtain livery accommodation there; if not, the owners should be able to recommend suitable yards. It is also worth asking knowledgeable people in your area, such as British Horse Society representatives, Pony Club committee members and instructors and staff in tack and feed shops.

At present, livery yards in the UK do not need to be licensed, but that is likely to change under forthcoming animal welfare legislation. The British Horse Society operates a voluntary inspection scheme, under which yards must have public liability insurance, comply with the latest health and safety legislation and receive annual inspections without warning to make sure standards of care are kept to a good level. It is a good indicator of standards but does not necessarily mean that yards who do not belong to it have lower ones: it is important to visit potential yards, if necessary with someone more experienced, and form your own impressions.

A good yard manager will be happy to show you round by appointment, when you may also get the chance to talk to clients, but good yards often have waiting lists – if you find the perfect yard that has a vacancy, it is worth paying to reserve accommodation before you find a horse.

Assessing a yard

When you walk onto a yard, your first impressions should be that it is reasonably tidy and in good repair. It doesn't have to look as if it belongs in a lifestyle magazine – there is much more to a good yard than sparkling paint work and hanging baskets – but you would not want to see tools lying around where they could be trodden on, doors that do not shut properly and lots of children or dogs running around out of control. You should also be able to inspect stables, fields and fencing, which should all be suitable and in good order, as explained in the next chapter.

A happy atmosphere, which you should be able to gauge by talking to the owner and perhaps clients, will lead to relaxed and happy horses. There should be evidence of security measures, especially if you intend to leave your tack in the yard tack room, and also of fire precautions. It is a good sign if the yard owner or manager asks you lots of questions and you should ask about rules for livery owners, which should be on display.

Questions to ask and things to work out include:

- How far away is the yard and how easy or difficult will the journey be at peak times?
- Is there good off-road riding or roads suitable for hacking?
- Will your horse be able to go out in the field every day? Some yards do not permit turnout in winter, which is not fair to your horse.
- Are the facilities adequate for what you want to do with your horse? For instance, if you want to ride on winter evenings, you will need access to a floodlit schooling area or an indoor school.

- Does someone keep an eye on the yard at night? If there is no one within sight or hearing overnight, it compromises security and also your horse's welfare.
- If you are looking after your horse, can the yard supply you with feed and bedding at reasonable rates or will you have to buy it elsewhere? Some people may in any case prefer to do this. If you are going to follow this route, is there adequate storage space?
- Do you intend to buy a horsebox or trailer? If so, and you do not have room to park it at home, is there a suitable place to park it where it will not be an open invitation to thieves?
- Can you fit in with the routine of the yard? For instance, it may be stipulated that all stables must be mucked out by a certain time each morning.
- Are there routine tasks which are shared between clients, such as sweeping the yard and removing droppings from the field?
- What insurance cover does the yard owner have? Public liability is essential.
- Does the owner require you to have insurance? Again, public liability is important and some yards will insist that your horse is insured against veterinary fees – a sensible move in any case and one which is explained fully in Chapter 13.
- What is the policy on worming? Some yard owners will want to worm all horses themselves and charge owners accordingly, whilst others will specify a programme and leave it up to individuals to buy and administer the appropriate worm medication. If there is no policy on worming, your horse's health may be at risk (see Chapter 09).
- Does the yard owner give you a written contract stating responsibilities on either side and rules of the yard? If not, problems and misunderstandings are more likely to arise. It is important that there should be a written contract between you and the yard owner detailing costs and payment intervals, the amount of notice that must be given on each side and responsibilities of each party.

In this chapter you will learn:

- **about field suitability, safety and maintenance**
- **how to choose fenching and shelters**
- **how to assess stable design and construction**
- **how to choose bedding materials and muck out.**

The best environment and lifestyle for a horse is to provide him with suitable grazing, amenable company of his own kind and shelter from bad weather or from flies and biting insects in milder conditions. Much of his time will be spent in the field, so it is important that this is safe and provides suitable grazing.

Field facts

As a guideline, you need a minimum of one acre per horse to keep the field in good condition all year round. Grass generally has more nutritional value in spring, summer and autumn – and in some cases, grazing may need to be restricted because too much of a good thing can causes weight and health problems – but has little food value in winter. As climate patterns change, so does grass growth and in a mild winter, it may keep growing and provide more nutritional value than expected. Grass needs a combination of warmth and moisture to thrive, so may become parched and stop growing during periods of summer drought.

Ideally, a field will have been sown with a mixture of grasses suitable for horses. If it was originally grown for dairy cattle it will be richer than a horse needs and you may have to restrict his grazing – not necessarily by keeping him out of the field, but by restricting his grazing to a smaller area fenced off with temporary electric fencing (called strip grazing). You may have to do this in any case for ponies susceptible to laminitis, a crippling foot condition now thought to be connected mainly to types of carbohydrates found in grass when it is of a particularly high feed value.

Horses should always have access to clean water. This can be provided in the field by a water trough, sited either along a fence line or well away from it – so there is no risk of a horse getting trapped between the trough and the fence – or in a suitable container with no sharp edges. All troughs and containers should be checked and cleaned out regularly.

Safety factors

Safe fencing is vital and there are several suitable types, including wooden post and rails and permanent electric fencing. Thick hedging is ideal, as it provides shelter as well as a boundary, but in many cases hedging needs to be reinforced with post and rail or electric fencing to make it secure enough. Plain wire is sometimes seen but is not recommended and barbed wire should never be used for horses as it can cause terrible injuries.

Gates should also be safe and well balanced, so they are easy to open and close. They must be at least 1.8 m wide to allow a horse to pass through safely but in most cases will need to be much wider to allow tractors and field maintenance equipment to pass through. Gates can be made from wood or galvanized metal and must be kept in good condition.

Fields should be checked every day so that broken fencing can be repaired and other hazards spotted and put right. Rabbit holes should be filled in immediately, as a horse may tread in one and break his leg. Unfortunately, you may also have to remove rubbish thrown into the field by passers-by; this can range from grass clippings to discarded drink cans. Horses find grass clippings palatable but must not be allowed to eat them, as they soon ferment and can cause colic. Other garden rubbish such as clippings from trees and shrubs may be poisonous.

Poisonous plants

One of the biggest field dangers comes from poisonous plants, in particular ragwort, which can cause fatal liver damage in horses. This is a biennial plant and destroying it takes at least a two-year programme – longer if the land has been infested for some time. Seeds can be dormant in the soil for up to 20 years. It forms rosettes in the first stage and yellow flowers in the second (see Plates 1 and 2). During the rosette stage, which happens after seeding, it lies flat to the ground about 7 cm across and is easily missed, so it is important to check for it regularly. The best way of removing it is to dig or pull it up, wearing gloves to protect yourself from the toxins it contains, then take it away from the field and burn it.

Other harmful plants include bracken, foxgloves and members of the nightshade family. Some trees may also be potentially harmful, especially yew. If there are any oak trees in or around a field where horses graze, they must be fenced off in autumn and winter as horses will eat acorns, which are poisonous to them.

Give him shelter

Horses in fields need shelter all year round. Although they will happily withstand much colder temperatures than we find bearable, they do not like a combination of wind and rain. In milder conditions, they need a refuge from flies and biting insects.

Trees and hedges may provide natural shelter, but in most cases horses will need a purpose-built one. This can be a conventionally designed wooden field shelter, which looks like a stable with one or more openings to allow horses to enter and exit at will, or an open-plan windbreak comprising solid sections about 2 m high, set like the spokes of a wheel. In most parts of the country you will need planning permission to site a field shelter, though there are mobile versions which can be towed by a suitable 4WD vehicle or tractor and are said to be exempt.

The size and siting of a shelter is important. It should be at least 5.5 × 3.5 m for two horses with plenty of room for them to go in and out without one feeling threatened by the other; the safest option is usually to leave the front completely open. Ideally, a shelter should be sited so the back is to the prevailing wind and on the highest part of the field so that water drains down and there is less risk of the ground becoming churned up by hooves.

Field maintenance

Fields need to be looked after all year round if they are to provide food and a safe environment. If you keep your horse on a livery yard, hopefully the owner will keep them in good condition, but if you have your own land or rent grazing on a DIY basis you may find that you are expected to do the work yourself or pay someone else to do it. For fields of up to two acres, it is possible to carry out most of the work without heavy-duty agricultural machinery, but once you get over that size you will need specialist equipment. This usually means employing a local farmer or agricultural contractor. In all cases, it is a good idea to understand what needs to be done.

Unlike sheep and cattle, horses are selective grazers, which is a polite way of saying that they can be very frustrating in the way they graze some areas right down and leave others to grow long. If the field is not managed properly, this will leave areas of long, rough and unpalatable grass combined with parts which have been grazed right down.

The best way to encourage more even grazing is to pick up droppings as often as possible, preferably every day, and to keep the grass at a regular height during the growing season by topping it – cutting it so that it is at an even height. Horses usually designate certain parts of the field as 'toilet' areas and if droppings are not removed, will become indiscriminate. It is important to remove droppings in any case to minimize the worm burden. All animals have internal parasites whose eggs are expelled in droppings and then taken in again as they graze, but removing droppings is an important way to break the cycle.

When you are dealing with a small field, or there are several people to do the job, the simplest way is to pick up droppings in a wheelbarrow. Large yards may used motorized ride-on field 'vacuum sweepers' which suck up the piles through a large pipe and deposit them in a trailer. Some yards harrow droppings to spread them instead of removing them, but this is nowhere near as effective. To have any effect, it must be done in hot, sunny weather so that the sun's UV rays kill at least some of the eggs and larvae. Harrowing a field in wet weather merely spreads the problem over a wider area.

Topping a field to an overall height of about 7.5–10 cms encourages horses to graze more evenly and also promotes better growth, as regular cutting encourages the grass to 'tiller out' – grow more blades of grass on each stalk – thus producing a denser, more hard-wearing sward. If this produces a lot of clippings, they should be raked up and removed; not only could they cause digestive problems if eaten when they have fermented, but they will also bury the grass underneath and inhibit further growth. A ride-on tractor lawnmower will do the job in a small field, but for a larger one you will need a tractor with a heavy-duty topper.

Land that is grazed continually will become over-grazed. The best way to prevent this is to take the horses off it for regular periods so that the grass can recover. If more than one field is available, simply graze and rest them alternately; if you have just the one, split it into sections with temporary electric fencing so that one section can be grazed whilst the other rests.

If possible, it is a good idea to graze the field with sheep or cattle occasionally, as they will eat the rougher grass that horses tend to avoid. Sheep do less damage on wet ground than cattle, which is why they are sometimes referred to as having 'golden feet'!

Harrowing and rolling are other routine tasks that will benefit your grazing, but must be done during the right weather conditions. Harrowing grassland in the spring, as soon as the land is dry enough to allow vehicles on it without causing damage, will pull out dead grass and leave room for new growth. Rolling can help repair damage done by horses' hooves in wet conditions, but again should not be done when the land is so wet that the machinery simply causes more damage. Different types of land have different characteristics – for instance, sandy soil is different from heavy clay – so get advice from an agricultural specialist or local farmer who understands how your type of land needs to be managed.

Fertilizing should be done with care and only when needed; a field that is well managed all year round and large enough to support the number of horses grazed on it may not need fertilizing every year. Get specialist advice on what type of fertilizer to apply, and when, to avoid stimulating a sudden growth of lush grass that could spark off digestive problems and laminitis in susceptible animals.

Maintaining a healthy, dense growth of grass will discourage weeds, but some are persistent no matter how carefully the field is managed. In these cases it will be necessary to use a herbicide; again, get specialist advice, because there are regulations governing their use and they must only be applied in suitable weather conditions. With some chemicals it may be necessary to remove the horses from the field for a period after application. Small areas can often be treated using a knapsack sprayer but agricultural machinery will be needed for large fields.

Stable thoughts

A stable should provide an environment in which your horse can be healthy and content. To achieve this, you need to pay careful thought to its design and also to the bedding you put in and the way you manage it. Some horses are happy to be stabled and will be all too ready to come in when the weather is bad, whilst others dislike being confined and may become distressed, showing stereotypical behaviour such as weaving, crib biting, box walking and wind sucking (see Chapter 11).

At one time these were known as stable vices, but it is generally accepted that this is unfair terminology and that the behaviour is due to stress. A horse who weaves will move his head from side to side over the door; in bad cases, he may move his whole body and shift his weight from one leg to the other. A crib biter gets hold of a convenient surface, usually the top of a door or a ledge, and bites down on it. Wind sucking means that the horse gulps down lungfuls of air; crib biting and wind sucking often, though not always, go together. Box walking is, as the name suggests, when a horse walks round and round his stable.

These behaviour patterns are a sign of mental stress but may also trigger physical problems. Weaving and box walking put stress on the limbs whilst crib biting and wind sucking may cause digestive problems and colic. Horses rarely weave outside and obviously don't box walk, so the best answer is to keep these animals outside and only stable them when it is absolutely essential. Some crib biters will continue the behaviour outside by crib biting on a fence or gatepost, but usually to a lesser degree.

Horses who are compatible with their neighbours will be much happier if the top half of the dividing wall between two stables is replaced by a metal grille which allows them to see each other and touch noses. An extra window opening, as explained below, will also be appreciated by many horses.

There has also been increasing interest in and support for stable mirrors – mirrors made from special safe, toughened material which are fastened to one of the walls. Many horses apparently take comfort from seeing their own reflection, though researchers do not seem sure if they recognize it as such or think it is another horse!

Size and structure

There are two sorts of stabling arrangements, internal and external. Internal stables are constructed within a large building, usually with stables on each side and a central walkway. This arrangement is appreciated by those looking after them in bad weather but it can be harder to provide good ventilation, which is important.

External stables are individual buildings, often constructed in rows. They are sometimes called loose boxes, a term which was coined to differentiate them from the old-fashioned stalls in which the occupants were tied up and able to lie down, but not to turn round – as opposed to being loose.

A stable should be as large as possible, no matter what the size of its occupant, but there are minimum dimensions. It is usually recommended that a stable should be at least 3 m × 3 m for a pony up to 14.2 hh; 3.6 m × 3.6 m for a horse up to 16 hh and 4.2 m × 4.2 m for a horse over 16 hh. However, these are the bare minimum and all horses will appreciate having as much space as possible.

Stables can be built from brick, concrete blocks or wood – timber which has been treated to withstand the weather. These days, wooden stables are the most common, mainly because fewer planning authorities are prepared to give planning permission for materials that could be looked on as permanent. This is a reaction to cases in residential areas where brick stables were built and then quickly turned into accommodation for humans.

The stable base is as important as the building itself. Concrete is an almost universal choice, as it is easy to keep clean; it can be laid with a roughened surface to help prevent the horse slipping and sloped slightly to allow urine to drain into a suitable outlet. Its only disadvantage is that it is cold, but this can be overcome by laying rubber matting designed especially for stables on top.

The roof should be pitched rather than flat, to allow the maximum volume of air, and internal headroom should be at least 3.6 m – more for a tall horse. The stable walls should be lined with 'kickboards' that minimize damage from flying hooves to the overall structure and also provide insulation. Doors openings should be at least 1.2 m wide so there is no risk of the horse banging himself as he goes in or out. They must open outwards, for safety reasons: horses sometimes roll over and get stuck against a wall (called getting cast) and if a door opened inwards it could be impossible to get in and release a trapped animal.

Stable doors traditionally have two halves, but the top half is largely dispensable. It should not be closed even in bad weather, as this stops airflow and creates a huge hazard to the horse's respiratory system. If you are worried about a horse being cold, it is better to put on a warmer rug. Just about the only time you would need to close the top door would be if a vet was testing the horse's eyesight and needed a dark environment in which to do it.

Bottom doors should have fastenings at top and bottom, for security: some horses learn how to open a top fastening with their teeth and one who kicks the door may cause enough movement in a bottom one to release it. The best system is a sliding bolt at the top, preferably with a covering to slide the end into so the horse can't get hold of it, and a kickover latch at the bottom that you can open and close with your foot – more convenient than having to bend down to unfasten a second bolt, especially if you are carrying something.

Many stables have only one window opening, usually in the front next to the door, but ventilation can be greatly improved by adding another window at the back. As this also gives the horse an extra view, he will often appreciate it. If necessary, the rear window can be closed, but this should only be necessary if a combination of driving rain and wind means the bedding will get soaked. It is important to differentiate between good ventilation, which means frequent air exchanges inside the building, and draughts, which a horse will not appreciate.

Air vents set into the stable greatly improve its ventilation qualities. Standard guidelines are that, in external stables, there should be inlets just above the horse's height and outlets in the roof ridge. This optimizes air circulation, as the air which comes in is warmed by the horse's body heat before rising and exiting through the roof. Internal stabling should be designed with specialist advice to make sure that ventilation is tailored to the design of the individual building.

Fixtures and fittings

Many stables will have internal lighting, which is a real bonus on dark winter mornings and evenings. It is important that all electrical installations are horse and rodent proof – cables should run through pipes which can't be chewed by either and switches should be housed in weatherproof casings and set outside, out of reach of the horse.

Other fittings may range from mangers to brackets set into the wall so that buckets of water can be held in them. Some yards use automatic drinkers, plumbed-in systems with a shallow drinking bowl which fills automatically as the horse drinks. These are convenient from a labour point of view, but mean that you can't tell how much your horse is drinking, so he may become dehydrated before you realize.

Most stables have tying-up rings set into an internal wall as well as outside. It is useful to have an internal ring for the odd occasion when it may be needed, but avoid mucking out a stable with a horse inside, or grooming him inside, as this subjects him to the effects of dust, dirt and ammonia and other toxic fumes from soiled bedding.

A home at home

Most horse owners, no matter how good their livery yard, dream of keeping their horse at home. If you are able to make this a reality, you will nearly always find that you need planning permission to build stables and other facilities, such as riding arenas.

There are many good companies who specialize in manufacturing and erecting stables and, although they will not apply for planning permission on your behalf, they will supply you with photographs and diagrams to back up your application. You will find that in most cases, manufacturers of external stables will expect you to have a base constructed to their specifications; this is a skilled job and best suited to a professional builder.

If you have a building which you think might be suitable for housing internal stables – which again, usually requires planning permission – specialist manufacturers will make site visits and advise you accordingly.

Bedding time

There are many types of bedding material, from the traditional straw and dust-extracted wood shavings to rubber matting, shredded cardboard, dried hemp and even elephant grass. Although your choice will be affected by cost and availability, the most important thing is that it should contain minimum dust and mould spores, both harmful to a horse's respiration system.

Horses are not nesting animals and lie down happily in the field, but need bedding on top of a concrete stable floor to provide warmth and secure footing and to reduce concussion. Rubber matting used alone will answer all these requirements, but some horses are reluctant to urinate on it because they do not like getting splashed – and as droppings get kicked about, your horse and/or his rugs will get very dirty. Rubber matting with bedding on top is an excellent option, but with most systems the mats must be lifted regularly and the floor washed through.

Widely available bedding materials and their pros and cons include:

- Wheat straw – cheap and usually easy to obtain and a feasible option as long as it is of good quality, not dirty or dusty. All straw contains some dust, which makes it unsuitable for horses with any degree of dust allergy. Some horses will eat it, which can cause digestive problems and perhaps colic, as straw contains an indigestible woody fibre called lignin. Because it is not wrapped, straw must be stored carefully, as if it gets wet it will become mouldy.

- Dust-extracted straw – chopped into short lengths and packed into wrapped, compressed bales, this is a healthier option but may not be as hard-wearing.
- Wood shavings – bales of compressed wood shavings are widely available, but must be dust-extracted. Shavings sold for use in poultry sheds will not have had the dust taken out and so should not be used for horses. The wrapped bales are easy to store and handle.
- Shredded paper and cardboard – these are completely dust and mould free to start with but, inevitably, dust from the environment will find its way in. They are heavy to handle when wet and because they do not rot down quickly, it may be more difficult to dispose of the resulting manure.
- Hemp and elephant grass – as these, like straw, are organic products, their quality may vary. Some manufacturers advise that the bedding should be sprayed with water to 'settle' it, but as moulds thrive in damp environments this cannot be recommended.
- Wood fibre products – these may be more absorbent than ordinary wood shavings, and may also be quicker to rot down on a muck heap, but they are usually more expensive.

Mucking out

A stable should be cleaned – mucked out – every day and droppings taken out as frequently as possible whilst the horse is in there. Once a week, or according to the manufacturer's instructions, use a disinfectant product recommended for use in stables, either in liquid or powder form.

Starting a new bed in an empty stable will take at least four bales and possibly as many as six or eight, depending on the floor space and the bedding material used. If you are putting the bed down on a concrete floor, it should be deep enough so that the prongs of a fork do not touch the floor when stuck into it. Some people like to use less on rubber matting, as the matting provides warmth and minimizes concussion, but matting and a reasonably deep bed often make the perfect combination.

Making the bed deeper at the sides, called banking, will help minimize draughts when the horse is lying down.

To muck out, you will need a range of tools: a wheelbarrow, a suitable fork, a shovel with a plastic head, and a yard broom with stiff bristles. A four-pronged fork is best for straw, whilst shavings and other products are easier to deal with if you have a lightweight shavings fork designed especially for the job. Some people prefer to wear stout rubber gloves and pick up droppings by hand – which is by no means as unpleasant as you might imagine!

With a straw bed, remove any droppings and sort the dry, clean straw from that which is wet and dirty. Pile the dry straw at the sides of the stable and put the dirty bedding in the wheelbarrow to go on the muck heap. When the dirty straw has been removed, sweep the floor and, unless the horse has to go straight back in the stable, leave the bedding piled at the sides to allow the floor to dry.

To muck out a bed made from shavings or other material, remove the droppings and scrape back the top layer of bedding to uncover any wet patches. Take out the wet bedding and replace with dry.

At one time, it was common to use a system called deep litter bedding, when droppings were removed daily but wet bedding was left and covered with a fresh, dry supply. At intervals ranging from a month to six months, the box would be fully mucked out and the dry bedding used to form the basis of a new bed. However, now that we know the harmful effects ammonia and other fumes from urine can have, as well as the fact that damp bedding provides an environment in which mould spores will thrive, deep litter systems are no longer recommended.

Protect yourself

You can't muck out a stable without throwing dust into the air, so tie the horse up outside when you are doing it rather than tying him up inside. Anyone who suffers from asthma or who is mucking out several stables each day should think about wearing a lightweight dust mask, available from any agricultural supplier or DIY shop.

Lifting bales of bedding and bags of feed and pushing wheelbarrows of soiled bedding will help to keep you fit – and owning a horse is a lot more fun than going to the gym – but can also put strain on your back if done incorrectly. The most important way to protect your back is to split heavy loads into smaller ones, where possible, and use the correct lifting technique: stand as close as possible to the load, bend your knees and keep your back straight whilst you lift.

As you lift, pull in your stomach muscles to increase intra-abdominal pressure. If possible, lift in stages, from one level to the next. When carrying loads, keep your back straight rather than stooping. Don't hoist heavy hay nets or bags of feed over your shoulder, as stooping causes more pressure and therefore wear and tear on the spine.

In this chapter you will learn:

- **how a horse's digestive system works**
- **the rules of feeding**
- **how to assess a horse's condition**
- **about types of feed and forage.**

Horses live to eat – not because they are greedy, but because that is the way they have evolved. Their digestive system means that they are 'trickle feeders' who need to eat small amounts at frequent intervals rather than large meals, infrequently. When given a natural lifestyle, with plenty of time in the field, they will spend most of it grazing: researchers have found that given the chance, a horse will graze for up to 15 hours out of 24.

Horses' lifestyles, including their eating patterns, are controlled by their owners, so it is important to mimic nature to satisfy both their physical and mental health. This means that when a horse is not in the field, he needs forage, such as hay, to keep him occupied and to keep his digestive system working smoothly. This, plus a permanent supply of clean, fresh water, will supply his basic needs.

When grass is at its most nutritious, it will supply most and sometimes all of a horse's needs. If it is particularly rich and he is not working hard, it may even provide too much of a good thing and you will have to reduce his grass intake by strip grazing, as explained in Chapter 03. This applies mainly to ponies, but also to horses who naturally gain maximum benefit from their food; they are often referred to as 'good doers' and often have a high proportion of draught horse or pony blood.

Other horses, especially those with active working lives, will need extra fuel from compound feed formulated especially for horses in the form of cubes or muesli-type mixes, often known as hard feed. At one time horses were fed 'straight' cereals, mainly oats and barley, but although some people still feed them, they have largely been superseded by commercially made feeds, consisting of cereals and other ingredients, and which include vitamins and minerals.

How the digestive system works

Digesting a mouthful of grass starts when the horse bites it off, then chews it. Saliva, which lubricates the food and also acts as a buffer to acid in the stomach, is produced only during the chewing process, not in anticipation of it – *your* mouth might water in anticipation of a nice meal, but your horse's won't.

The chewed food passes into the stomach, then into the small intestine. Food that is not digested here – which will be mostly fibre – goes into the large intestine. The large intestine is often compared to a huge fermentation vat, as it contains micro-organisms which break down fibre and extract nutrients.

Any food from which nothing useful can be extracted is passed out of the body.

The golden rules

With so many types of feed available, it can seem a confusing subject. Following some basic principles will help you keep it relatively simple.

- Always feed plenty of forage, of which the main types are grass, hay and haylage. Hay is dried grass and haylage is grass which has been cut, wilted and packed in sealed bags or wrappings to preserve a higher moisture content and nutritional value. In nearly all cases – and just about the only exceptions are racehorses and those competing at top level in three-day eventing – forage should make up 75 to 100 per cent of the total diet and many horses will work happily and stay healthy on a regime of 100 per cent quality forage.
- Feed only the best quality. For instance, this means clean, well-made hay that is as free from dust as possible and shows no sign of mould, and feed made by a reputable specialist manufacturer.
- Always make sure your horse has a supply of clean, fresh water.
- Aim to maintain your horse in correct condition, neither too fat nor too thin. The scientific way of assessing this is to use a system known as condition scoring, explained later in this chapter.
- Give hard feed on the basis of small meals at regular intervals. Do not give more than 2 kg hard feed at a time or you will overload the horse's system.
- If you make changes to the diet, perhaps by switching from one brand of feed to another or starting a new batch of hay, do it gradually over several days by mixing in a little bit of the new with the old and gradually altering the proportions. This will keep the digestive system healthy.
- Increase the horse's exercise before you increase his feed, not the other way round.
- Horses like routine, so if he gets hard feed, try and give meals at the same times each day. If you keep your horse on a DIY livery yard, you may be asked to prepare and leave your horse's morning feed the night before so that all horses can be fed at the same time.

- Feed by weight, not by volume. A scoop (a round plastic dish with a handle generally used to dispense feed) of one type of feed will not necessarily weigh the same as a scoop of another, so always weigh a new type of feed before using it; every now and then, weigh a scoop full to check that you are not inadvertently feeding too much or too little.
- Store feed and forage so that it stays clean and dry and can't be contaminated by vermin such as rats and mice.

Weighty matters

Obesity can be just as much a problem with horses as it can with people; there are far more horses who are too fat than are too thin. The generally used scale of condition scoring, expressed as 0–5, with 0 being so thin the horse's welfare is at stake and 5 being obese – also a welfare consideration – may help you to assess a horse. You should aim for a condition score of 3.

0 Very poor. The rump is very sunken and there is a deep cavity under the tail. The skin looks as if it is tight over the skeleton and the backbone, ribs and pelvis are all prominent. The horse will look as if it has a ewe neck (upside down) even when this is not actually the case.

1 Poor. All the above apply, but to a slightly lesser extent.

2 Moderate. The neck is firm but narrow and the rump is flat either side of the spine, though the spine itself is not visible. Ribs can usually be seen, though not markedly so as with the first two categories.

3 Good. The ribs can be felt but are covered, the rump is rounded and the neck is firm without being cresty.

4 Fat. There is a gutter along the back and rump and it is difficult to feel the ribs and pelvis.

5 Obese. The gutter along the back and rump is deep, the ribs cannot be felt and there are pads of fat – usually on the crest, body and shoulders.

How much does he need?

Although some people never bother to find out what their horse weighs, it is a really good idea to do this. Not only will it enable you to work out the quantities he needs more accurately, it will be useful when you have to calculate quantities of worming medication. Regular bodyweight checks will also enable you to keep your horse's weight standard; when you see him every day, it is sometimes difficult to spot gradual changes.

Obviously you can't buy a pair of heavy-duty bathroom scales, though veterinary practices and some racing yards have special weighbridges that do the same job. For the ordinary owner, the easiest method is to use a weightape, a special tape measure which allows you to measure your horse's girth and read off a corresponding weight. These are available from many saddlers, feed merchants and feed manufacturers and although they are not totally accurate, they will show when a horse is gaining or losing weight.

When calculating quantities, it is generally accepted that a horse needs to eat 2–2.5 per cent of his body weight in total each day, so a 500-kg horse would need 12.5 kg per day. Most horses, other than those in really hard work, are best suited to a ratio of 90–100 per cent forage with 0–10 per cent of the diet comprising concentrates (often called hard feed) if necessary. If he needed to lose or gain weight, this would be done by altering the food value rather than the quantity – so assuming that our 500-kg horse is on a ratio of 90 per cent forage and 10 per cent concentrates he would need 11.25 kg forage and 1.25 kg hard feed.

Grass has a high water content and will provide between half and all of a horse's forage needs, depending on the time he spends grazing. When he is stabled, make sure he has suitable hay or haylage to keep him happy.

Although all owners should understand the principles of feeding, don't give yourself a headache over the figures. The well-known feed companies all employ nutritionists who will give free advice and help you formulate a feed plan – obviously they will only recommend their own brand, but if you talk to more than one, you should find that the advice remains the same and only the brand names change.

Forage facts

Many owners will be able to buy good quality hay or haylage through their livery yard, but others will need to buy in from outside sources. In either case, you need to be able to recognize the difference between good and not so good. Always buy them from a reputable grower or merchant, as it is vital to make sure that hay does not contain ragwort; it is hard to distinguish individual ragwort plants once hay has been baled and it is more palatable when it is dried.

The feed value of hay depends on multiple factors, including the grass mixture, its maturity at cutting and how well it was made. Lucerne hay, made from lucerne, or alfalfa, has a high feed value and should only be fed to horses in really hard work, such as racehorses.

Although forage can be analyzed to identify its feed value, this is not usually necessary for the average owner. Hay which is not dusty when shaken out, has a sweetish smell and is green rather than brown will usually be of good quality. However, as all hay contains a small amount of dust, nutritionists advise that it should now always be soaked.

Soaking hay in clean water for a short time – from about 15 minutes for enough for one helping for one horse to several hours for a complete densely packed bale – minimizes the risk from dust spores. The hay should be completely submerged, which will probably mean weighing it down in the soaking container, and you should use fresh water for every batch soaked.

At one time, it was advised that hay should be soaked overnight. Today's nutritionists say this is unnecessary and lowers the feed value of the hay, as nutrients are destroyed.

Haylage is more convenient, because it does not need soaking and – unlike hay, which loses some of its feed value with age – has a guaranteed nutritional value. It is more expensive, though, and must be made by a producer experienced in horses' needs. If something goes wrong whilst it is being made, or a bag or wrapping is punctured, haylage can go mouldy. If this happens, discard the whole bale; a good supplier will take back any mouldy bales in exchange for new ones.

Silage, which is fed mainly to dairy cattle, is also made from grass baled before it has dried. However, additives may be put in that are not suitable for horses and it will usually be too rich for them. Even more important, it is made in a way that means there is a risk of it containing micro-organisms that cause botulism, which does not affect cows but can be fatal to horses.

Forage is also available in the form of short chopped feed, marketed under various brand names. Chaff, traditionally short chopped hay or hay and oat straw, is also sold under brand names. Some forage feeds, such as those made from dried, chopped alfalfa, have quite a high feed value and so should not be fed in large quantities without careful thought.

Supplying extra fuel

If your horse needs extra fuel to work, this can be supplied in the form of commercially made compounds (hard feed). These contain a variety of ingredients with added vitamins and minerals, made into either mixes or cubes (also called pellets or nuts) and formulated to different nutritional levels. Cubes are usually cheaper than mixes, because they are cheaper to make, but although mixes appear more pleasing to the human eye there is no difference in feed value between cubes and mixes of the same nutritional level. Most horses will happily eat either, though some who are fussy feeders may prefer mixes.

It is important to feed a product that has been formulated to suit your individual horse and the work he is doing, not what you hope to do. For instance, a feed designed for a hard-working competition horse will not be suitable for one in light work and will only make him too fat. It may also affect his behaviour, as you will be supplying more energy than he needs. As a guide, work levels have been defined by nutritionist Ruth Bishop of Spillers Horse Feeds as:

- Maintenance – horses and ponies not in work.
- Light work – hacking and leisure riding, 1–2 hours per day; showing at local level; Prelim/Novice to Elementary level dressage; show jumping at local shows; BSJA British Novice and Discovery classes; endurance rides up to 30 km at a slow pace, including sponsored rides; unaffiliated one-day events at novice level.
- Medium work – showing on the affiliated circuit; affiliated dressage at Elementary to Medium and Advanced Medium; BSJA Newcomers and Foxhunters level; BE intermediate and advanced one-day events; 80-km endurance rides; fast canter work for racing.
- Hard work – affiliated dressage at Grand Prix level; BE three and four star three-day events; endurance race rides; racehorses in full training and racing.

Most feeds are packed in bags printed with feeding quantity guidelines. Because these have to cater for a huge range of animals they are often over generous, so don't assume that your horse will need the full amount. Horses usually find hard feed palatable, so to slow down their eating rate it is a good idea to mix in a handful of chaff and to dampen the feed slightly.

Other feeds

Some people still prefer to feed straight cereals in conjunction with chaff and perhaps soaked sugar beet. This usually means oats or barley that have been rolled or crushed, though maize is popular in the USA. Oats are a traditional feed for horses and cause few digestive problems. They have a reputation for causing horses and, in particular, ponies to become more lively, hence the expression 'feeling his oats'!

It is now thought that this is due not so much to the energy level, which is actually lower than barley, but because the energy in oats is released quickly. As many people ride their horses soon after feeding them, this seems to fit. A relatively new form of oat called Naked Oats, which is fed whole, gets round the problem; it has a higher oil content and the energy it provides is released more slowly.

Although oats and barley are incorporated in many commercial feeds, they are usually cooked at very high temperatures to make them more digestible. When fed alone, their main drawback is that they have a nutritional imbalance – their poor ratio of calcium to phosphorus has to be compensated for. This can be done by feeding a commercial oat balancer, which is just as suitable for feeding with barley, or alfalfa, which is high in calcium.

Soaked sugar beet is a very useful feed. It is actually made from what is left of the sugar beet after the crop has been made into refined sugar for the human food industry and usually comes in the form of pellets or shreds. Some types must be soaked for several hours, but there are also new formulations that are safe to feed after soaking for several minutes. It is vital to read the instructions on the bag, as if it is not soaked for the minimum recommended time it will absorb liquid from the digestive system and continue to swell, setting up a high risk of colic.

Sugar beet is known as a succulent. Other succulents include apples and carrots, which most horses love. Apples should be cut in quarters and carrots sliced lengthwise so there is no danger of them causing the horse to choke.

Feed balancers have become popular over the last few years. They supply high levels of nutrients and are fed in very small quantities, which makes them useful for fussy feeders. They can be fed alone, with a forage feed or chaff, or with a reduced amount of hard feed.

Supplementary benefits

There is a huge number of special feed supplements for horses, just as there is for humans. These are widely advertised and marketed – and, it has to be said, the advertising and marketing is so successful that many owners buy them when they don't need to! If your horse has a balanced diet and no particular problems, he does not need any routine additives apart from salt, either in the form of a free-access salt lick or 50 g of common salt daily added to his feed.

However, if he may be lacking in nutrients or has a particular problem, there may be times when he will benefit from nutritional support. It can't be emphasized too strongly that if you think your horse has a problem, you should always start by getting veterinary advice. For instance, if you think your horse is moving stiffly, get him checked out rather than automatically buying a feed additive said to help in cases of stiffness and arthritis. If he does have arthritis, then your vet may advise that nutritional support is a good idea, but correct diagnosis is vital.

Additives that may at times be useful include:

- Broad spectrum vitamin and mineral supplements. If your horse is a good doer and maintains his correct weight on forage alone or forage and a minimal amount of hard feed, you may want to feed a broad spectrum vitamin and mineral supplement to compensate for any deficiencies. Although feeds are formulated with added vitamins and minerals, these are only delivered at the right level if you feed a minimum amount daily – usually 2–3 kg, which will be more than many animals need.
- Probiotics. These help to maintain a healthy environment in the gut. This environment can be disrupted if the horse is stressed – perhaps because he is moved to a different yard – or if he is given antibiotics.
- Garlic. Although its benefits are anecdotal rather than scientifically proven, garlic is said to be absorbed into the horse's system and leave a smell on the skin and coat that deters flies and biting insects. It is also said to benefit the respiratory system.
- Supplements to target particular problems, ranging from poor hoof quality to overexitability. But before you look for an answer in a tub, look at your horse's overall diet, lifestyle and management. Are his hooves weak and crumbling because his overall diet is deficient? Is he excitable because he

does not get enough time in the field, or because his feed is supplying more energy than he needs?

Water matters

Water is as important to your horse's health and well-being as feed. It makes up 60 per cent of his body weight and, in hot conditions, he will drink up to 70 litres in 24 hours. Dehydration is as dangerous for horses as it is for people, which is why one of the golden rules of feeding is that clean, fresh water must be available at all times.

Horses are fastidious about the quality of the water they drink, so it is important to ensure that the water itself and the container is clean. In a stable, water will absorb the taint of ammonia from urine, so must be changed daily. Automatic watering systems keep the supply fresh, as long as the drinking bowl is kept clean, but the drawback is that you cannot monitor how much the horse is drinking.

Many people put drinking water in buckets or other containers on the floor and, whilst most horses are happy with this in familiar surroundings, research has shown that some animals are much happier to drink if their water is placed off the ground, either in corner mangers or from buckets in wall clips. Another method, which suits some horses, is to put their water in a manger which hangs on the outside of the stable door, so they can see what is going on around them whilst they drink. It is thought that some horses are reluctant to drink from floor level because they feel vulnerable.

It is equally important to make sure that clean water is always available in the field. Troughs and containers should be cleaned out regularly to prevent a build-up of algae, and checked twice daily to make sure that refill mechanisms are working properly and that there are no 'foreign bodies' such as leaves or dead birds in the supply. In freezing conditions, ice should be broken several times a day.

How to feed

As a grazing animal, the horse's natural feeding position is with his head on the floor. Whenever possible, it is best to mimic this by feeding from the floor; this means that as he chews the food, his teeth meet at the correct angle. Hay and haylage can be placed directly on the floor or in special containers, whilst feed can be put in tough containers – ones which are designed to fit inside an old car tyre usually deter the horse who likes to kick his feed bucket round the stable.

If it isn't possible to feed hay from the floor, you may want to follow the traditional method of feeding it from a haynet. This should be tied to a ring mounted in the stable at a height suitable for the individual horse; it should be high enough that there is no risk of him getting his foot caught in an empty net, but not so high that he is holding his neck at an awkward angle to eat and there is a risk of seeds falling in his eyes.

To tie up a filled haynet, pull the drawstring round the top tight and pass it through the tying-up ring or loop. Take the end down the side of the haynet, slot it through the net near the bottom and pull it back up, tying it in a quick release knot between the ring or loop and the top of the haynet, passing the free end through the loop you have made. This means that as the net empties, it does not drop down so that there is a risk of the horse putting his foot through it and getting caught up.

As was discussed in the previous chapter, it is important to try and keep the stabled horse happy. One of the best ways is to make sure that he always has a supply of forage, so he can at least approximate his natural grazing behaviour.

Another way to make him work for his food, and to make a small amount of feed last longer, is to use feeding 'toys'. These are containers with a small hole, designed to be filled with cubes. The idea is that the horse pushes the container around the stable so that a few cubes at a time fall out.

Storage

Feed and forage must be kept clean, dry and as safe as possible from vermin. Disease can be spread by rats, mice and birds. Hay should be stored on pallets to let air circulate, ideally in a weatherproof building. If this is not available, keep it covered with one of the modern tarpaulin covers, made from waterproof, breathable material. These are much better than the old-fashioned tarpaulins, which cause condensation and encourage the growth of mould spores in the resulting damp environment.

Galvanized feed bins are the ideal storage medium for hard feed, but are expensive. Many owners who have just one or two horses keep their feed in dustbins bought especially for the purpose – but use metal ones, not plastic, as rats can chew through plastic dustbins with surprising ease.

When you finish one lot of feed, make sure the container is clean and dry before you put in another, as old food will go stale and affect the new batch.

05 groomed for success

In this chapter you will learn:

- **why grooming is important**
- **what equipment is needed**
- **about grooming techniques**
- **how to bath your horse.**

The value of grooming

Daily grooming is an important part of routine horse care. It helps keep a horse smart – but even more important, it helps keep him healthy. Grooming allows you to spend time with him, notice his behaviour and be alert for any problems, such as a reaction to pressure that may signify discomfort, or heat in a limb that may be an early warning of potential lameness. It also helps maintain a healthy skin and coat by keeping the pores of the skin open, removing waste products such as dried sweat and ensuring that there is no dirt to get trapped under tack and rugs and cause rubs. Realistically, some owners may find it difficult to find time to groom thoroughly every day, but it is important to pick out feet, remove the worst of any dried mud and dirt and carry out a basic routine.

Passive stretching exercises, which can be incorporated into a grooming routine, can be used to warm up muscles before exercise. You don't have to do them every day, but two or three times a week will help your horse, just as a warm-up routine helps human athletes – including riders!

Brushing basics

The natural grease in a horse's coat, which looks like light coloured dust, acts as a waterproofing agent. This, together with a winter coat that is longer and thicker than his summer one, protects against bad weather. Thorough grooming removes grease and dirt, so grooming regimes need to be tailored according to a horse's lifestyle. One who lives out all or most of the time will need a simpler routine than one who is stabled for longer periods and clipped during winter.

The grass-kept horse needs his feet picking out and dried mud brushed off, particularly from any areas where it could cause rubs, such as under tack. His eyes, nostrils and dock (the area under the tail) should be kept clean but the grease in his coat should not be removed by vigorous grooming. This is especially important if he does not wear an outdoor rug.

With horses who are stabled part of the time, follow the full grooming routine outlined later in this chapter.

Essential kit

A grooming kit should contain the following items.

- A hoof pick is used to remove packed dirt, droppings and foreign bodies such as stones. It is useful to keep a folding hoof pick in your pocket when out riding in case of emergency.
- A rubber curry comb is used to remove dried mud, dried sweat and loose hair and brings dirt and grease to the surface of the coat. It also has a massaging effect, which many horses enjoy.

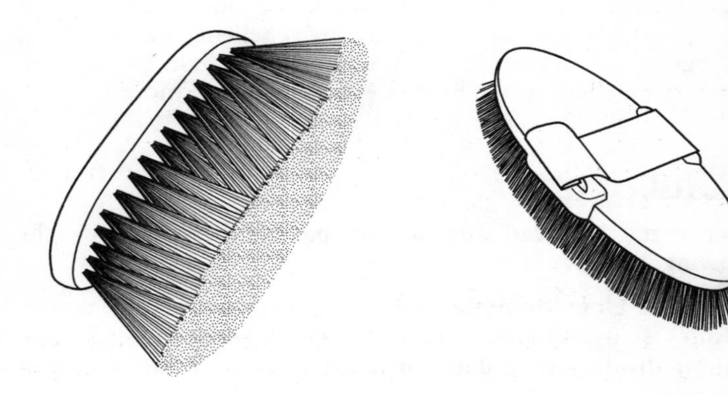

figure 3 dandy brush and body brush

- A dandy brush has long bristles. Traditional designs with stiff bristles are used to brush off dried mud and sweat from the body but should only be used on thick coats and always with care. They are too harsh for horses with fine coats and more sensitive skin, who will find them uncomfortable. They should not be used on manes and tails as they will break the hair. Dandy brushes with softer bristles, sometimes called whisk or flick brushes, are gentle enough to use on most horses. They are used to flick away dirt on the surface of the coat.
- A body brush has short, soft bristles and can be used all over, including on the head area. Its main use is to remove grease from the coat and it is gentle enough to use on the mane and tail, as it will separate hairs without breaking them.

- A metal curry comb is only used to clean the body brush and should never be used on the horse itself. The body brush is drawn over the teeth to bring out the grease and dirt trapped in the bristles and the curry comb tapped out to remove it.
- Small sponges or cotton wool pads can be used to clean the eyes, nostrils and dock, using a separate one for each area to avoid spreading infection. Cotton wool is more hygienic as it can be thrown away after each use. Use a separate sponge or pad for each eye.
- A stable rubber is a cloth made of loose weave material – a tea towel works well. It is used slightly dampened at the end of the grooming routine to remove the last traces of dust.
- Fly repellent is essential when flies and other insects are a nuisance. Allergies to commercial fly repellents are rare but it is a good idea to 'patch test' a new product first. Do not risk getting repellent in the horse's eyes, nostrils or mouth.

Useful extras

Other items you can add to the basics above include the following.

- A cactus cloth or mitten made from a very coarse woven cloth is useful for tackling sweat and stable stains. Alternatively, use a dampened facecloth and rub dry with a towel.
- A sweat scraper is used to remove excess water after a horse has been bathed or rinsed off. The safest designs are made from plastic, with rubber blades.
- A metal hair shedder is useful for removing dead winter coat; the commonest design has a curved blade with short, fine metal teeth. It must be used with great care.
- A human hairbrush is used by many professional grooms on manes and tails.
- Hoof oil or dressing, usually applied with a brush, is used to create a smart appearance for shows and competitions. Ask your farrier's advice about what to use, as some products may have a drying effect if used too frequently.
- Mane and tail combs made from metal or plastic are used to separate mane hair into sections when plaiting. Pulling combs, with short, narrowly spaced teeth, are used when pulling – thinning and/or shortening – manes and tails.
- Face brushes are small brushes with soft bristles, often made from goats' hair, which can be used on the faces of sensitive horses.

Safety and health

Grooming a horse puts you close to a heavy animal who could easily hurt you without meaning to: as most of us find out, when a horse steps on your foot, your toes come off worst! This means that he should always be tied up safely and that you need to be careful how you position yourself (see Chapter 01). You are also dispersing dirt into the atmosphere, so need to protect both his and your respiratory systems.

Whenever possible, groom a horse outside so that you are not releasing dirt and dust into the confined area of his stable, where breathing it in will not do either of you much good. If you are asthmatic, you may need to persuade someone else to do your grooming or at least wear a face mask, available from most DIY shops and agricultural merchants.

Always tie up a horse safely to groom him. Make sure you tie him to a proper tying-up ring screwed into a suitable site, such as a wall; there should be a loop of strong but breakable string between the rope and the ring so that if he is startled and pulls back, the string will break in an emergency. Always use a quick release knot, as explained in Chapter 01. Don't tie him to a gate or a fence rail, because if he is startled and pulls back, they may not hold even when a safety loop is used, and you could end up with a frightened horse trying to run away from a broken rail attached to his headcollar rope. As an alternative to using a safety loop, you can buy headcollar ropes with breakaway attachments designed to give way under a certain amount of force.

If your horse is ticklish, be considerate and don't use brushes with stiff bristles. Some horses dislike being brushed on the belly, head or leg areas, so be careful how you position yourself and be prepared to use a soft brush or even the flat of your hand. Face the hindquarters as you groom and stand close to his body, using the hand closest to it – hold grooming implements in your left hand when grooming the near (left) side and your right hand when grooming the off (right) side. This enables you to be more effective, as you can put your body weight behind the strokes of the brush when necessary. Never kneel down beside or stand directly behind a horse, and be aware of his reactions as you work.

Grooming techniques

Thoroughly grooming a partly stabled horse who is in work will take about half an hour, longer if you are really dedicated. Afterwards, you should feel as if you have made a real effort. Grooming is more effective when the horse is warm, as the pores of his skin are more open, so the ideal routine – if you have time – is to give him a quick check and brush over before you ride and groom thoroughly when you return.

Before you tack up, pick out his feet, run your hands over his body and legs to check for any lumps, signs of heat or swelling and brush off the worst of any dirt, especially where the saddle or bridle rests. If you don't have time to groom thoroughly after exercise, pick out the feet again and brush off dried sweat or sponge the area with lukewarm water as appropriate.

The routine below can be followed when a horse does not need to wear a rug. In cold weather, when you are working with a clipped or fine-coated horse who needs to be kept warm, unfasten and fold the front half of his rug back over his loins whilst you groom the front half of his body on each side, then fold the back half to the withers whilst you groom the back half.

Step 1 Clean out his feet with your hoof pick – if you always work in the order of near fore, near hind, off fore, off hind, the horse will learn to anticipate and if he is really obliging, will pick up the appropriate foot even before you ask. Don't grab at his legs, but put the hand nearest his body on his shoulder or hindquarters and run it gently but firmly down to the base of the limb until you reach the fetlock. A gentle squeeze should encourage him to lift his foot so you can support it and use the hoof pick.

Clear any dirt or embedded stones, working in a downwards action from heel to toe so you do not dig the end into his foot. Clean down the sides of the sensitive V-shaped frog but be careful not to dig inside it. Whilst you are cleaning out his feet, check his shoes to make sure they are secure.

Step 2 Starting at the neck, use the rubber curry comb in circular movements to lift dried mud, dirt and grease. Work downwards to the shoulder and gently down the forelegs, then go back to the withers and work along and down the body and hindquarters. Be equally careful as you deal with the hindlegs and belly: rubber curry combs can be used gently or with more pressure as appropriate and most horses prefer them to stiff dandy brushes. Horses are all individuals, though, so be prepared to experiment, with care. Follow the same procedure on the other side.

Step 3 A whisk or flick brush is ideal for the next step. Starting at the neck and following the same paths as before, use quick strokes ending with a quick turn of the wrist to flick the dust and dirt raised by your rubber curry comb away from the coat. Repeat on the other side.

Step 4 Use a body brush in short, firm strokes to remove deeper grease and dirt and massage the skin. Put the weight of your body behind your arm; this is the original 'elbow grease'! Every few strokes, draw your brush over a metal curry comb to remove dirt from the bristles and tap the curry comb on a suitable surface to release it. To clean the brush effectively, you need to move it over the curry comb, not vice versa; pushing the brush away from your body as you do this rather than pulling it towards you means you will not release dirt over your hands and clothes.

Most horses do not object to the gentle use of a body brush on the head area, but with those who do, it is better to use a small face brush, a soft cloth or your hand.

Step 5 Use the body brush or hairbrush and/or your fingers to separate mane and tail hair. To brush out a tail, stand at the side of the horse and hold the tail halfway down. You can then brush through the end section of the tail, gradually moving up its length and brushing down. This way, you will remove any tangles rather than create them. Use the same technique on the mane and forelock. Some owners like to use commercial mane and tail conditioners, usually sold in spray form, every few days on the lower half of the tail to deter tangling. These make the hair slippery so be careful not to get them on the saddle area.

Step 6 Using dampened sponges or cotton wool pads, gently wipe the eyes, nostrils and lips and dock.

Step 7 Wipe over the face and body with a dampened stable rubber to remove last traces of dust. Your horse should now look and feel clean and comfortable whilst you should feel that you have put in some hard work!

Step 8 If necessary, apply fly repellent. If your horse dislikes being sprayed, a technique that often works is to get a helper to hold him, then brush the coat and spray immediately afterwards, alternating brush strokes with sprays. If this does not work, spray repellent on to a cloth and wipe it over him. Don't spray the head area in case you get repellent in his eyes; spray a small amount on to your hands and wipe round the eye and ears areas, then wash your hands. You can also buy cream or roll-on repellents, which may be easier to apply.

Stretches

There are many stretching exercises designed for specific purposes, but the two 'carrot stretches' explained here will benefit most horses. They should be performed within a range of movement the horse finds comfortable; don't ask for more than he is happy to give.

Carrot stretches are easy to perform and horses soon get the hang of them. Start by holding your horse in a headcollar, on a loose rope; then hold a carrot or other titbit against his shoulder, and encourage him to stretch his neck round to reach it. He will soon get the idea and, by moving the reward, you can encourage him to stretch round to his hip. This exercise should be carried out twice on each side and will help increase the flexibility of his neck and shoulders.

The second carrot stretch, which will loosen up his back and neck muscles, is carried out in a similar way but, this time, take the reward down and between his front legs.

Bathing

Bathing a horse is a convenient way to get rid of dirt, dust and dried sweat, especially when appearances are important. However, it might not always be advisable: bathing removes grease from the coat, which acts as a natural 'waterproofing' agent for horses who live out, so they should only be given an occasional full bath in periods of warm, dry weather.

It is also important to use the right shampoo – choose one formulated for horses, as it should not strip the skin of natural oils. Some horse owners like to use shampoo made for people, but this doesn't work out any cheaper than using a good-quality horse shampoo. Another golden rule is to rinse until there is not a trace of shampoo residue left in the coat, as this can lead to irritation and leave a coating on the hair that makes it dull instead of clean and shiny.

Shampoo for horses has become as high tech as shampoo for people. Some products contain ingredients said to act as fly repellent, whilst others are designed to bring out highlights in particular colours such as bay and chestnut. Traditionally, the old grooms would use a 'blue bag' in the final rinse for grey horses to bring out the brightness in white hair; now there are shampoos formulated to do the same thing.

Getting under the skin

The thickness of a horse's skin varies. It is thickest at the roots of the mane and the upper surface of the dock and thicker on the upper body than the lower part. Coat thickness varies between types and breeds: native ponies and horses with a lot of pony or draught blood usually have denser coats than those with a lot of Thoroughbred or Arab blood.

This means that whilst you can wash a mane or tail in less than ideal conditions, as long as you are careful to keep the rest of the horse warm, it isn't fair to bath a horse in cold or windy conditions. Grey, piebald and skewbald horses with pink skin, or those with white legs or muzzles, also need to be treated with extra care. The lack of pigment makes pale skin more sensitive and it may also be more susceptible to detergents – so whilst it is never a good idea to use washing-up liquid to bath a horse, it is even more important to avoid using them on pink skin.

In hot weather or after hard work, a horse will sweat under the saddle and girth. Washing off these areas with plain, warm water will make him comfortable; if necessary, put on a rug designed to help him dry off afterwards. If you need to remove grass or stable stains and water alone won't work, spot washing with shampoo or one of the many stain removers on the market can avoid the need for a complete bath.

Bathing techniques

Bathing a horse in hot weather should still be done with care and consideration. Before you start, use a rubber curry comb to loosen the dust and grease in his coat and get everything ready: a bucket of water in which shampoo is diluted according to the manufacturer's instructions, two sponges, a plastic sweat scraper for removing excess water, clean towels and a rug to help him dry off, if necessary. It is possible to buy rugs which transfer the moisture from a wet or sweating horse from the coat to the outside of the fabric, whilst keeping him warm and dry underneath.

Hopefully he will already be used to having water applied via a hosepipe, as this is the most efficient way of rinsing him off. If you are not sure, treat him as you would a horse who is unfamiliar with the experience, as some may be nervous. Turn on the water to give a slow trickle, then start by letting it run on to the ground near his feet. As he gets used to it, run the water over his foot and gradually work your way up the leg, shoulder and body. Increase the water flow as he gains confidence.

If it is not possible to use a hosepipe, you will have to use buckets of clean water. You will need several and it's easier if you can persuade a helper to keep you supplied. In any case, it is better to use a bucket of water to wet the coat on his neck and head, as few horses take kindly to having water sprayed down their ear from a misdirected hose.

The logical way to bath a horse is to start at the neck, including the mane, then go down the shoulders and front legs, along the body and down the hindquarters and hindlegs. Then wash the tail and finally go back to the head.

Unless the instructions on the shampoo state otherwise, start by wetting the coat with clean water, then apply the shampoo. Pay particular attention to the base of the mane and tail, as grease deposits collect here. When you get to the tail, use a separate bucket of clean water and wet it thoroughly; most horses will let you dip the long tail hairs direct into the bucket, but be cautious. Use your second sponge – for hygiene reasons, it's best to use a different one from that used on the head and body – to wet the top of the tail and rub in shampoo. Finally, use your original sponge to clean the horse's head carefully, making sure not to get shampoo in the eyes. You may not even need to use shampoo – plain water may be enough.

Rinse the horse from front to back, using either a hosepipe or buckets of water. As you rinse, rub the coat with your hand to work out all traces of shampoo. When the water runs clear, use the sweat scraper in the direction of the coat to remove as much water as possible, being careful not to press hard over bony areas. Then rinse and scrape again and repeat the rinsing process until you are sure there is no shampoo left in the coat.

The tail can also be rinsed off with either a hose or a bucket of water, depending on the horse. Not surprisingly, many horses don't like cold water being sprayed over delicate parts of their anatomy, so if you don't have the luxury of a mixer tap to supply warm water, it's more considerate and safer to use a bucket. Squeeze your hand down the long tail hairs to take out excess water, then stand to one side of the horse, hold the tail at the end of the dock (where the bones finish) and swish the hairs round in a circle to get rid of as much water as possible. If you're not sure whether the horse is used to this, it's a good idea to try a 'dry run' first – if he takes time to get used to you swishing his tail round when it's dry, he's going to be even less impressed if you introduce the experience when his tail is wet and has water flying off it!

On a hot day, your horse may dry off quickly if you walk him round. If you feel there is the slightest risk of him getting chilled, play safe and use a rug designed to help him dry off.

If it is essential to bath a horse in cooler conditions, you can give him the equivalent of a bed bath. Get everything ready beforehand, including a scraper, towels, a rug designed to help him dry off and, if appropriate, stable bandage or leg wraps.

Use warm water and start by washing and rinsing the tail, because you don't need to remove rugs to do this. Next, fold back the rug from the front to wash the neck and mane. Scrape, rinse and scrape again and towel dry.

Now fold the rug on to the loins and hindquarters and do the middle section, finally folding it forward to wash and dry the back end. Check that your horse is warm enough and, if not, add another rug before cleaning his head and legs and, if necessary, use bandages or leg wraps for warmth.

If it really is too cold for a bath, there is a useful technique that works well on horses with fine coats or those that have been clipped. Dip a piece of clean towel in hand-hot water and wring out as much moisture as possible. Folding and moving your rug as before, rub the coat vigorously, rinsing the cloth regularly.

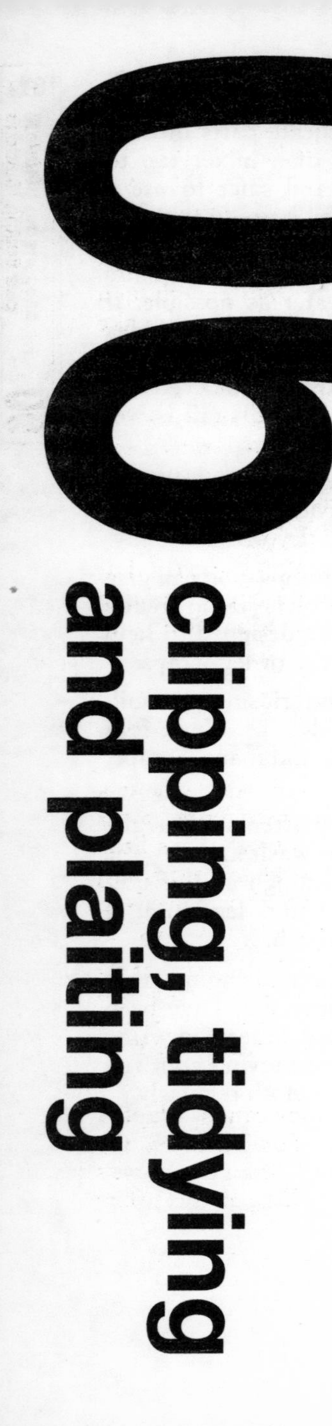

In this chapter you will learn:

- why, when and how to clip
- which sort of clip to choose
- how to tidy manes and tails
- how to plait manes and tails.

Clipping

Although the thickness of a horse's coat will depend on his breed or type, his winter coat will always be thicker than his summer one. If he is asked to do any more than very light work, he will sweat, and drying him off afterwards so he does not catch a chill will take a lot of care and time. Clipping off part – or occasionally, all – of his coat often makes it easier for him to work comfortably. It will also make him look smarter – though this should not be the main reason for doing it – and will make effective grooming of the partially stabled horse easier.

It is best to wait until your horse's winter coat is fully established before clipping for the first time. If he grows a relatively fine coat, as with Thoroughbreds, he may only need clipping once or twice during the winter. Thick, dense coats tend to re-grow more quickly and this type of horse may need several clips.

At one time the tradition was not to clip after January 1st, because it was thought this would spoil the summer coat coming through. However, nowadays most people clip according to need rather than the time of year.

Types of clip

There are several different types of clip, from ones where a minimal amount of coat is removed to those where more hair is taken off than is left on. The best policy is to take off as little coat as is needed to enable the horse to work comfortably, because the bigger the clip, the more protection a horse needs in the form of rugs and stabling. Clipping just a small area, in particular the underside of the neck and chest (see 'bib clip', below) will often make a big difference. If in doubt, leave it on: you can always extend the clip, but if you decide you have taken off too much of the horse's coat you will have to wait for it to grow back.

There are six generally recognized clips, but they can be adapted or combined to suit the individual horse. For instance, the bottom edges of a blanket clip can be high or low on the horse's sides. It is best to leave the hair on the horse's legs, as this offers valuable protection.

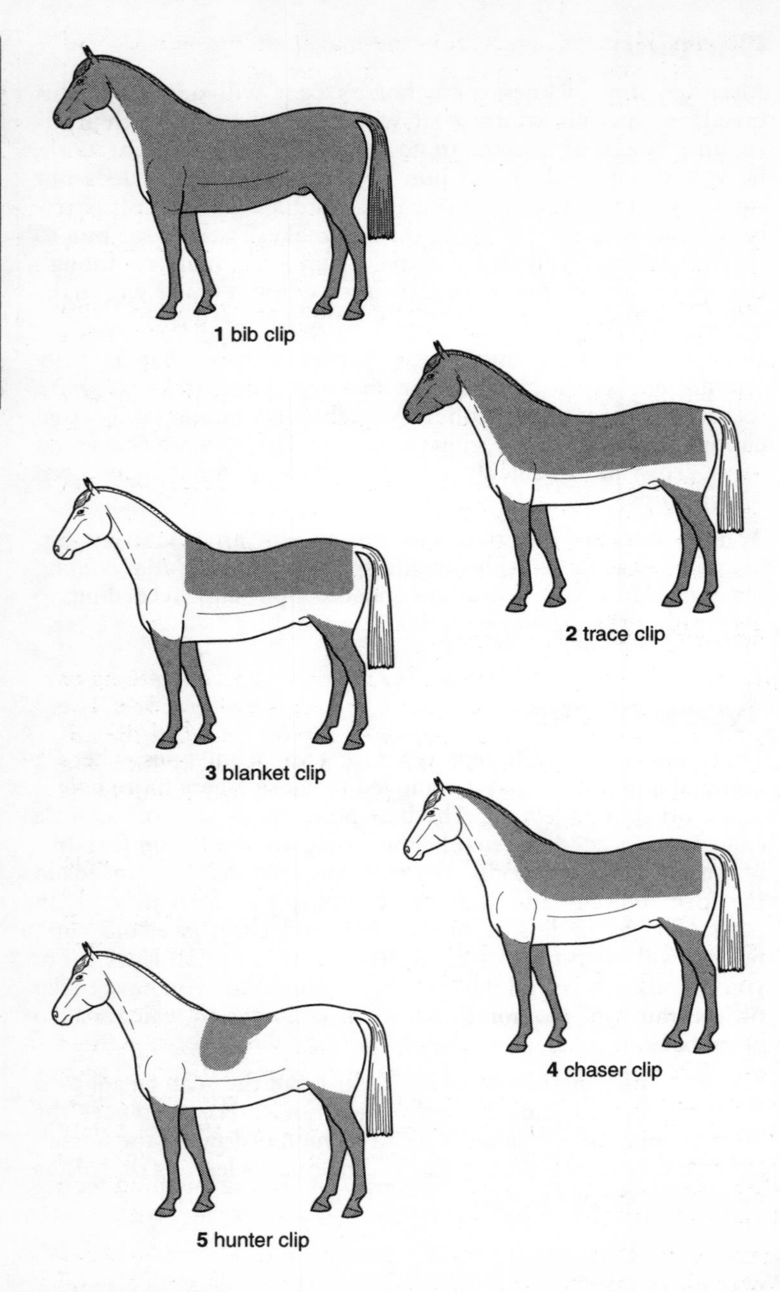

figure 4 types of clip

- **Bib clip:** Hair is clipped from the underside of the neck and from the chest. A good clip for a horse who lives out and needs natural protection from the weather, but who sweats if all his coat is left on. Also useful for introducing a young horse to the idea of being clipped, as it is quick and easy to do.
- **Trace clip:** Gets its name because it follows the lines of traces on driving harness. Another useful clip which can be used for a horse who lives out and wears a weatherproof rug.
- **Blanket clip:** Hair is left in a blanket shape over the back and hindquarters but clipped from the neck and all or part of the head.
- **Chaser clip:** Originally designed for racehorses, this clip is a variation of the blanket clip and leaves the hair on the upper part of the neck.
- **Hunter clip:** The only hair left on is that on the legs and the saddle patch – the area where the saddle sits.
- **Full clip:** The complete coat is clipped off, including from the legs. Drastic and not recommended for general use.

Clipping considerations

Successful clipping is a mixture of technique, correct handling of horses, the right clippers used in a safe way – and practice! The best way to learn is to watch someone experienced and then, if possible, clip a horse who is not worried by the procedure. If a horse has not been clipped before, it is best to ask someone experienced and considerate to accustom him to it.

The area where the horse stands to be clipped should be light and dry and out of the wind. If you clip him in his stable, pile up the bedding round the sides so the clipped hair can be swept out. Take out water containers and cover up automatic watering systems so that hair does not get into the water.

Clippers can be battery operated or powered by electricity. Battery operated ones are usually quieter, which makes them more likely to be accepted by an inexperienced or nervous horse, but may not be powerful enough to cope with large clips. Electric clippers should be used with due regard for safety precautions – for instance, extension leads should not be placed where they can be trodden on by horse or human and you should wear rubber-soled boots. A circuit breaker is vital.

Make sure that the clippers are comfortable to handle. Powerful, heavy duty ones needed by large yards with several horses to be clipped are usually heavier than models which may be perfectly suitable for the owner with just one or two horses to clip each season. You also need to check that the blades are suitable; in most cases, you will need fine or medium ones rather than coarse. They will need to be sharpened professionally at regular intervals.

It is sensible to wear a hard hat, as a ticklish or nervous horse may kick when you are working on sensitive areas such as the belly. If you have long hair, tie it back so there is no risk of it getting caught in the clippers.

Before you start, read the clipper manufacturer's instructions and tension them correctly. You will need special clipper oil to lubricate them throughout the process, a rug to put on your horse afterwards and, if you are clipping near the tail area, a tail bandage to keep the tail hairs out of the way. Unless you are doing a full clip, it helps to mark out the lines you will be clipping up to on the coat with a dampened piece of chalk.

Clipping techniques

No matter how experienced the horse, you should run the clippers for a few seconds without touching him to let him get used to the noise. Start on a less sensitive area, such as the shoulder, if possible; place your hand on the horse and rest the running clippers against your hand to introduce him to the vibration they cause.

Clip against the lie of the coat and try and make long strokes rather than short ones. Overlapping each stroke with the next will give a better appearance and minimize the risk of unsightly 'tram lines'. Every now and then, stop to lubricate the blades with clipper oil and make sure they are not hot.

Difficult areas

Some areas are more fiddly than others. You will need a helper to lift each front leg and stretch it forward whilst you are clipping the girth and elbow area, to avoid catching wrinkles of skin with the blades.

Be careful round the ear area, as some horses are particularly sensitive here. Don't clip the hair from inside the ears, because it prevents seeds and other 'foreign bodies' falling in.

Some people clip the whole head, but unless both clipper and horse are experienced it is safer to clip up to a line drawn from the base of the ears down to the corner of the mouth and leave the hair on the front of the face. Use a piece of string to find the ear to mouth line and chalk it in.

Opinions vary on whether or not whiskers should be clipped off. Some say it gives a finished picture whilst others believe they should be left on, as they act as a sensory aid.

Aftercare

A clipped horse will often be more lively at first, as he will be more sensitive to temperature. Keep him warm with rugs whenever necessary; you may need a special exercise rug that keeps his back and loins warm when riding slow work, such as hacking. Be careful not to keep him standing around without a rug.

Clippers need looking after, too. Clean and check them every time after use according to the manufacturer's instructions.

Manes and tails

All horses need their manes and tails shortening and tidying regularly, as otherwise they grow to an unmanageable length. Some breeds, such as Arabs and native ponies, are traditionally kept with flowing manes and tails – but even these may need tidying sometimes.

To shorten a tail, first watch your horse being led round in walk to judge the angle at which he naturally carries it. Then ask a helper to grasp the tail near the top and hold it so it is in this natural carrying position. Hold the end hairs together and shorten them by cutting across either with scissors or using a pair of clippers. Cutting whilst the tail is at the correct angle means the ends will hang level when the horse moves; if you cut the hair whilst the horse is standing still with his tail relaxed, you will get a sloping bottom edge instead of a straight one when he moves.

There are two good ways to shorten a mane that does not need thinning. One is to use an old clipper blade: backcomb the mane and use the blade to nip off the ends of the bottom layer you expose. You can also buy special combs with blades inserted at the base of the teeth to achieve the same effect.

Another way is to use scissors, but never cut straight across the bottom of the mane or you will end up with the equine version of the pudding basin haircut! Instead, comb a section of mane and nip off the ends of a few hairs with the scissor blades held at an angle, then comb and snip, comb and snip all the way up the neck. It is important to keep combing so you change the angle of the hair all the time and don't get 'steps' in the mane.

The traditional way to thin and shape a mane or tail is to use a technique called pulling, when you literally pull out a few hairs at a time. Some horses don't mind this, but it should always be done with consideration. Only do a little at a time and only pull when the horse is warm – usually after he has been worked – as the pores of the skin are more open and the hairs come out easily.

To pull a mane, you will need a pulling comb, which is a metal comb with short teeth. Use it to backcomb the top layers of the mane so you expose the hairs underneath, then wrap five or six round the comb and pull out in a quick motion. Only pull from underneath, because if you pull the top hairs, they will grow back in a spiky fringe along the crest of the neck.

A similar technique is used to pull a tail, but this takes more skill and is best learned by watching someone experienced. Take hairs from each side to reduce the bulk and work down until you are about three quarters of the way down the dock. Pull only a few at a time or you will be left with bald patches. Safety is vital, especially when pulling a tail. Never stand directly behind the horse, always to one side.

If a horse dislikes the pulling process – and some undoubtedly find it uncomfortable – it is not worth persevering. It is not fair on him and you are putting yourself at risk, as you are likely to get kicked. The easy alternative is to use a comb designed originally for professional dog groomers wanting to remove dead undercoat. It has a small head incorporating a curved clipper blade and by combing it down the sides of the tail you can achieve the same appearance as if it was pulled. It can also be used to thin out a mane, by combing down the underneath layer.

To keep a pulled or shaped tail tidy, you need to put on a tail bandage regularly so that the hairs lie flat. This should also be done when travelling a horse. Tail bandages are slightly elasticated and fasten with tapes or Velcro/hook-and-eye fabric.

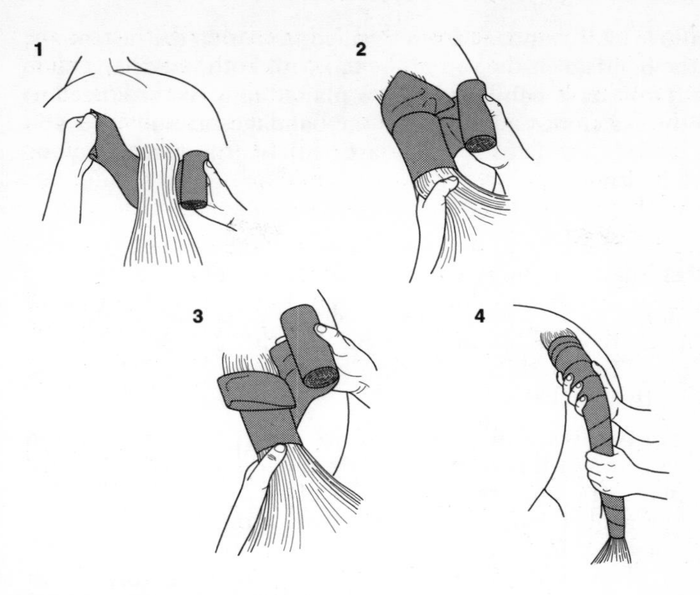

figure 5 putting on a tail bandage

Before you put one on, it should be wound into a tight roll with the fastener on the inside, so that when it is in place the fastener will be on the outside: if it is wound the wrong way, you will have to take if off and start again.

Dampen the tail hair before you apply the bandage. Don't dampen the bandage, or it will tighten as it dries and, if left on for a long period, could restrict the circulation and cause some of the hair to fall out.

Standing behind and slightly to one side of the horse, so there is no risk of you getting kicked, unroll 15–20 cm of bandage and put it under the tail. Make your first turn, then fold over the loose end and bandage over it; this helps the prevent the bandage slipping down. Continue bandaging down the tail, overlapping the bottom third to one half of the previous turn each time until you get to the end of the tail bones.

Now bandage back up until you reach the end of the bandage and secure. The fastenings should be no tighter than the tension of the bandage and if you fold the bandage over them, it helps to prevent them coming undone. Finally, bend the tail gently into its normal position.

To remove a tail bandage from a pulled tail, undo the fastenings, hold the bandage at the top of the tail with both hands and slide it off. If you have bandaged over a plaited tail, you will need to undo the fastenings and unwrap the bandage carefully – if you pull it off as before you are likely to disturb your careful plaiting.

Plaiting

If you hope to take your horse to shows or competitions, you may need to plait his mane to fit in with accepted turnout principles. To do this well takes practice and the more you practise, the quicker you'll get.

There are two ways of securing plaits, either stitching them in place with tough thread or fastening them with rubber bands. Sewn plaits are smarter and stay in place better than banded ones, but rubber bands are useful when you have to plait in extra fast time. In both cases, you need to make sure that the mane is of even thickness throughout and preferably about 10–12.5 cm long. It should be reasonably clean, or you will see specks of grease in the partings between plaits, but not newly washed – squeaky clean hair is slippery and hard to keep hold of, so wash the mane a couple of days before you intend to plait.

Comb the mane through and divide it into equal sections; it's easier to keep them separate by putting a rubber band round each one. Traditionally, there were seven or nine plaits along the mane and one for the forelock, but these days it is accepted that you can alter the number to suit the horse. Dampening the top of each section with light hair gel – the sort sold for people – helps prevent short hairs at the base of the mane sticking up.

Divide the section into three equal parts and plait down, keeping it tight from the base. Using a large needle and knotted thread the same colour as the mane, secure the end, then turn the loose ends over and wrap the thread round to keep them out of the way. Push the needle through the underneath of the plait at the top so it doubles up, roll up the plait and secure with two or three stitches.

Try and keep the plaits the same size all the way down the neck and sitting at the same height on the neck, then plait the forelock in the same way. When you want to undo the plaits, use a dressmaker's stitch unpicker rather than scissors – this enables you to hook and cut the thread without the risk of cutting into the mane hairs.

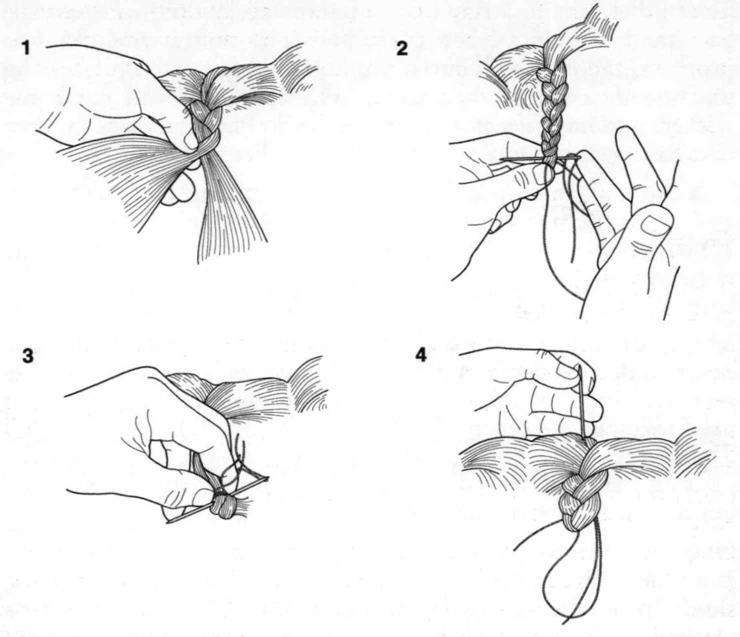

figure 6 plaiting a mane

To fasten plaits with rubber bands, plait down to the end of a section as before, then wrap the band round. Roll up the plait to the base of the neck and fasten with a second band.

Plaiting variations

The plaits described above are standard and acceptable for all disciplines, but there are other methods which can be useful.

Stable plaits are used to help train an unruly mane to lie flat. Simply divide the mane into sections, plait down and fasten the end with a rubber band. Leave the plaits in for a couple of days at a time and repeat as needed.

A running plait is a practical way of keeping a mane that is too long and flowing to be put into conventional plaits out of the way – for instance, when jumping or doing fast work. It is sometimes called an Arab plait, because it is often used on Arab racehorses.

Comb through the mane, then take a section of hair near the ears as if you were going to form a standard plait. Plait down as before, but take in a tiny piece of mane every time you pass the left hand section of the plait over the centre one. As you progress, the plait will curve round, forming a solid plait along the bottom edge of the mane. Eventually you will reach the withers and have no more hair to take in. Plait the hairs that are left, fold over the loose ends and stitch them securely.

You can plait the top of a tail instead of pulling it or shaping it, as long as the hairs at the dock are long enough to start it off. Take a small section, about 5 mm wide, from each side of the tail right at the top, cross them over and take a third section from the side – it doesn't matter which one – to the centre.

This gives you the three sections to form your plait. As you plait down, take extra hairs from each side as you pass the side sections over the central one – it is the same technique as that used to make a French plait, so if in doubt, ask a hairdresser! The secret of getting a really smart plaited tail is to keep the plaiting tight, so the side sections lie even and the central plait lies down the centre of the tail.

Once you reach about two thirds of the way down the dock, continue plaiting without taking in any more hairs from the side. This will give you a long, single plait that should be secured at the end with a needle and thread, doubled up and stitched in place.

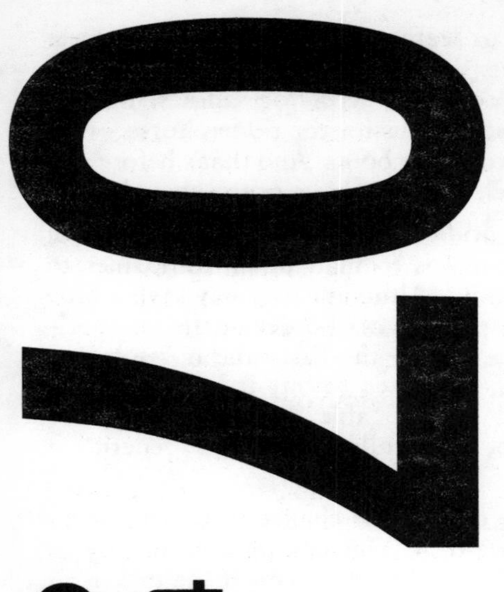

tack and equipment

In this chapter you will learn:

- **what tack and equipment you will need**
- **how to fit saddles, bridles and bits**
- **how to clean tack and equipment**
- **how to make sure it's safe and secure.**

Look in any saddlery store or search any equestrian internet site and you will see that owning a horse invariably means owning an awful lot of tack – a collective term for saddles, bridles, headcollars and other essential items for the ridden horse – plus other equipment, such as protective boots. And that's before you even think about rugs, which are the subject of the next chapter.

Whether you are buying a bridle or a rug, you will find lots of outlets, ranging from tack shops to mail order companies to internet sales sites and auctions. Although you may save a little on the purchase price by buying via mail order or the internet – though this isn't always the case – the first-time horse owner will find it is usually better, and safer, to buy from a reputable tack shop. Not only will you be able to see what you're getting, you should also, hopefully, have the benefit of experienced advice.

In some cases, you may be offered the chance to buy a horse's tack as part of the purchase deal. This can be a good way of saving money, but it can also lead to problems if the tack is of poor quality or doesn't fit. This applies particularly to saddles, so it is always a good idea for the novice owner to get someone experienced – preferably a professional who holds a recognized qualification, such as that from the Society of Master Saddlers in the UK – to check it for soundness, safety and fit before handing over any money.

It is as well to remember that the fact that someone is an experienced rider does not necessarily mean that he or she is an experienced saddle fitter. However, every rider needs to recognize the basics of saddle fit, and thus when to call in a professional to make adjustments, and how to adjust bridles etc., as explained in this chapter.

If you have to kit out your horse from scratch, it is always an investment to buy the best quality you can afford. Not only do your own and your horse's safety and comfort rely on it, it will last for many years if looked after properly and, on a cost per use basis, will usually end up giving you better value than tack which is initially cheaper but of inferior quality.

Good quality leather tack will be everyone's first choice, but not everyone can afford to buy everything a horse needs brand new. If you are working on a tight budget, you can save money by buying quality secondhand tack from a reputable retailer who has checked it and made any necessary repairs before putting it on sale. For instance, many good retailers take secondhand saddles in part exchange against new ones when customers sell a horse and need a different saddle for a new animal.

You may also want to consider synthetic tack, particularly in the case of saddles; there is a new generation of synthetic saddles which look very similar to leather. The latest designs, launched this year (2006) have not been available for long enough to know whether they will last as long as properly cared for leather saddles, but are much cheaper – and a good synthetic saddle is often better than a poor quality leather one. If you go down the secondhand or synthetic route, the same recommendations apply about professional fitting.

Sitting comfortably

There are different saddle designs for different purposes, ranging from the tiny, lightweight ones used by jockeys to jumping saddles designed for experienced riders tackling larger courses. As most people starting out in horse owning want to have a go at most things, from hacking to general schooling and jumping, the best type of saddle to start out with is a general purpose (GP) model. This will enable you to sit correctly and comfortably for most activities at a basic level.

Your saddle must fit you and your horse, with your horse getting top priority. A badly fitting saddle will cause discomfort and pain. Key points that an experienced person will help you to recognize are:

- Most saddles are built on a frame called a tree, which is made in different widths. It must be the correct width for your horse; if it is too narrow, it will pinch him and if too wide, it will come down too low on his back, causing equal problems. The tree must also follow the shape of the horse's back. If he has a flat back and your saddle has a curved tree, or vice versa, it will cause harmful pressure points.
- It should not restrict his movement when the rider is on board.
- It should spread the rider's weight over as wide an area as possible.
- The saddle should be level from front to back so that the rider can sit in balance, not be tipped forward or backward. A saddler/fitter will be able to make some adjustment and alteration.
- It should sit evenly, not over to one side. Again, a saddler/fitter will be able to tell if the problem is caused by an imbalance in the saddle, or by a rider putting more weight in one stirrup than the other.

- Someone standing behind the horse should see that the gullet (the channel down the centre of the underside) is clear of the horse's back throughout, including under the rider.
- The pommel at the front and the cantle at the back should clear the horse's back sufficiently.
- There is bound to be slight movement when you ride, but the saddle should not rock or bounce noticeably.

There are several types of treeless saddle on the market. They undoubtedly have a place, but have not yet won the same acceptance as those with trees.

To fit the rider comfortably, a saddle should have a large enough seat to accommodate the backside. The knee rolls – blocks at the front of the saddle underneath the flaps which give security – should be in the right place and the flaps should be the correct length to suit the dimensions of your legs.

Fixtures and fittings

As well as a saddle, you will need stirrup irons and leathers, a girth and perhaps a numnah or pad to go under the saddle if your saddle fitter advises it. As the girth holds the saddle in place and the leathers and stirrups are vital for your security and safety, don't compromise on quality.

Many riders now prefer stirrups of a safety design, such as the 'bent leg' ones with angled sides, designed to minimize the risk of a rider's foot being trapped in a fall. Safety stirrups with rubber rings on one side should only be used by small children, as the weight of larger youngsters and adults puts too much stress on the metal.

All stirrups should allow 1.25 cm clearance on each side of the widest part of the foot. There should be no more than this, or your foot could slide forward. Rubber treads on the base of the irons give cushioning and security.

There are many designs of numnahs and saddle pads, but unless your saddler recommends one for a particular purpose, you shouldn't need anything other than a thin cotton design – and then only to help absorb sweat and keep the saddle clean. It must be of a design that can be pulled up into the gullet – and will stay there – or it will press down on the withers and back.

Bits and bridles

A bit better

There are thousands of different types of bit, which can be divided into different groups: for reference, these are snaffles, double bridles, pelhams and gag snaffles. Bitless bridles make up a fifth group. In most cases, a horse suitable for a first-time owner should work nicely in a snaffle – though sometimes you may find you need a different bit, such as a pelham, in situations such as jumping or cantering in company.

It isn't that one bit is 'right' and another is 'wrong'. What is important is that you should be happy and in control and your horse should be comfortable. Bitting is a complex subject, so if you are not sure what type of bit to use, get advice from a good trainer who can watch you and your horse and advise accordingly. There are many factors to take into account, such as the rider's balance in the saddle, the horse's way of going and the shape of his mouth, lips and tongue.

Whatever bit you choose, it must be the right size and be correctly adjusted in the horse's mouth. Bits are measured from the inside of one cheekpiece to the other; some manufacturers work in inches and some in centimetres. If the bit has a jointed mouthpiece, it should be straightened before measuring.

When a bit is at the correct height it will fit snugly into the corners of the mouth with a wrinkle at the corner of the lips. If the horse's mouth is pulled into a fixed smile, it is too high, and if there is no wrinkle, it is probably too low. These are general guidelines because you have to take into account factors such as how fleshy the horse's lips are.

To judge the size, straighten a jointed mouthpiece in the mouth. There should be just enough room to fit your little finger – about 1 cm – between the bit cheekpiece and the horse's lips on each side.

If you use a pelham or kimblewick (a variation from the same bit family) the curb chain must be fitted so it cannot twist. With these particular bits, the easiest way to ensure this is to fasten the chain on the right-hand hook, twist it clockwise so it lies flat, then pass it through the kimblewick rings or the top ring of the pelham before passing it through the same ring on the left-hand side and hooking it on the left-hand hook.

Bridle basics

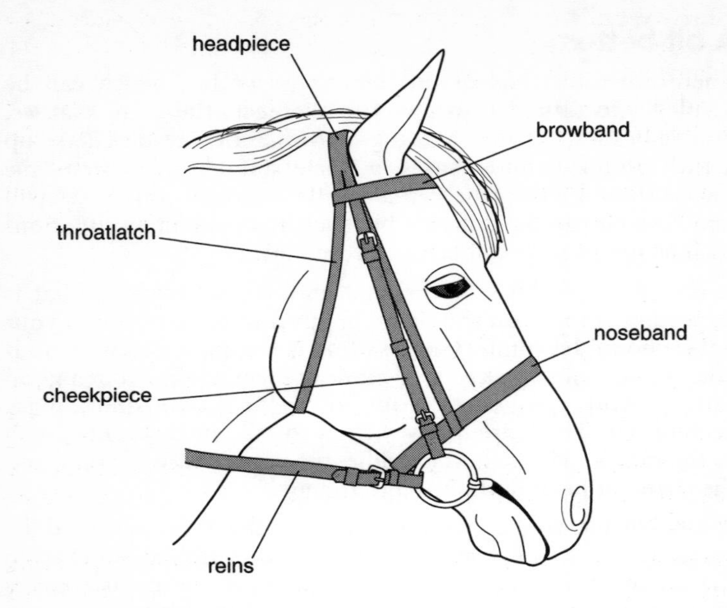

figure 7 parts of a bridle

Bridles are usually sold in Shetland, pony, cob and full sizes, but to get a good fit it's sometimes necessary to mix and match parts from different sizes. For instance, you may find that a native pony with a broad forehead may need a pony- or cob-size bridle with a browband from the next size up.

It's important to make sure your bridle fits well and is adjusted correctly. If it doesn't, your horse will be uncomfortable and will show it by resistances such as throwing his head up and down. The main fitting guidelines are:

- The browband should be long enough not to pull the headpiece forward so it pinches the ears. However, if it flops up and down it is too long.
- The throatlatch (which for some reason is pronounced throatlash) should be fastened so you can fit four fingers' width between it and the horse's cheek.

- Plain cavesson nosebands should not be so high that they rub the base of the cheekbones. Allow at least one finger's width between the noseband and the cheekbone on each side. They should be fastened so you can fit two fingers between the noseband and the horse's face.
- Some nosebands fasten above and below the bit and are designed to prevent the horse opening his mouth too wide. Keep this in mind rather than thinking of them being used to keep the mouth shut, as if they are too tight, the horse will not be able to flex his jaw and he will not be able to respond correctly to the action of the bit.

Headcollars and halters

Headcollars are used for leading a horse and when tying him up. They can be made from leather or webbing. In the USA, they are usually called halters, but in the UK, a halter is different – it is made from webbing or rope. Some designs are not suitable for tying up, because they could come off over the horse's ears if he pulled back.

Pressure headcollars and halters are specialist training equipment. They are used in conjunction with the handler's body language (see Chapter 01) to teach a horse to answer to light pressure. These should never be used when tying up or travelling a horse, as if he pulls back, the pressure will not release and he may panic or injure himself.

Lead ropes should be made from a material such as cotton rope that is strong, but does not burn your hands. Some materials can do this when a horse pulls away even if the handler is wearing gloves – which is always recommended when leading. The leadrope should be fastened so that the clip faces away from the horse's jaw; this means there is no risk of it digging in to him.

Martingales and breastplates

Many riders use martingales to help give more control when riding out and jumping. The two most commonly used are running and standing and all are designed to help prevent the horse putting his head too high. Running martingales work via the reins and standing martingales use the nose as a control point.

To fit a running martingale, stretch the straps which work on the reins along the underside of the neck and alter the adjustment at the girth so that the rings reach the horse's gullet. The other way is to stretch the rings back to the withers and alter the fit until there is a hand's width between the withers and the rings, but this does not allow for the fact that horses' shoulders slope at varying angles and is not always as accurate. If there is a kink in the reins when the rider holds them in the normal riding position, the martingale is too tight.

A standing martingale should be fitted so that it fits into the gullet when pushed up along the underside of the neck. It should only be fastened to nosebands which fit above the bit, never to ones that fasten below, or it will affect the horse's ability to breathe.

Breastplates and breastgirths, designed to help prevent a saddle slipping back, should be used when extra security is needed, not to try and compensate for a badly fitting saddle. They are often combined with martingales in one piece of equipment.

If your horse doesn't need a martingale, it's often a good idea to use a neckstrap – it will give you a bit of extra security if you need it. A spare stirrup leather buckled round the horse's neck will do.

Lungeing equipment

Most owners need to be able to lunge their horses sometimes, either because there are occasions when they are unable to ride or because they are experienced enough to use it as a way of improving the horse's schooling. Although lungeing equipment isn't usually something you need to buy immediately, it will be something you'll want to add to your wish list.

The basics for exercising an experienced horse on the lunge in an enclosed area are a lunge cavesson – which is like a headcollar with a reinforced noseband into which metal rings are set – a lunge rein and a lunge whip. Cavessons can be made from webbing or leather; leather ones are much more expensive and a well-made webbing one will be fine in most cases.

Lunge reins are made in several kinds of material, but cotton webbing is ideal. It is easy to hold and non-slip. The lunge whip should feel well-balanced and easy to handle, so you can point it at different parts of the horse's body depending on the action you are trying to encourage.

Tacking up

When you know a horse is quiet and safe, it's possible to tack him up in a stable without tying him up. But if you haven't got to know each other yet, or you are tacking him up outside, he must be tied up before you start.

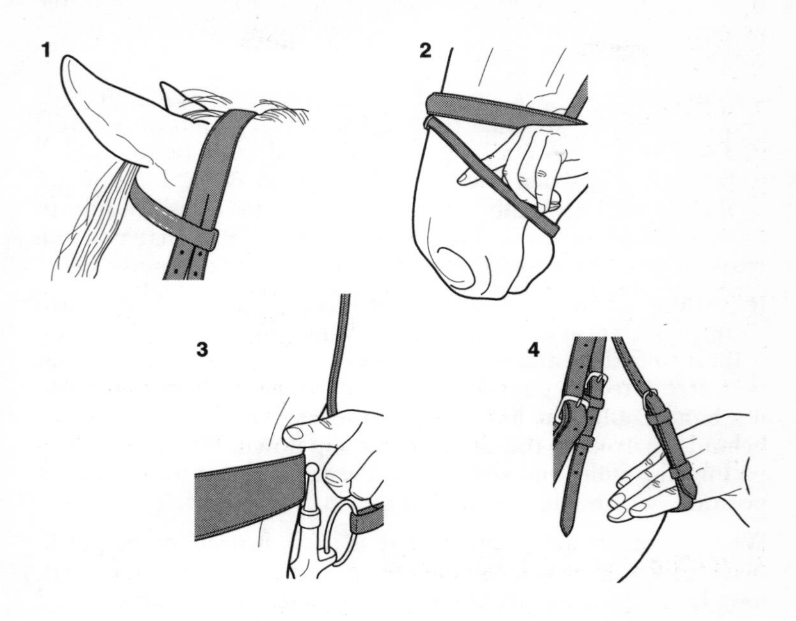

figure 8 correct fit for a bridle

To put on a bridle, pass the reins over his head, then undo the headcollar and fasten it round his neck. This gives you some control if he decides to walk off. Hold the bridle in one hand in front of the horse's face and support the bit with the other, guiding it gently into his mouth. Be careful not to bang it against his teeth, or he will throw up his head – and repeated clumsy handling will often make him object to having the bridle put on.

Some horses are very obliging and will open their mouths to take the bit. With others, you will need to slide your thumb into the side of the mouth and gently press down on the gap between the teeth of the lower jaw, where the bit rests.

Now pass the headpiece over one ear, then the other, handling the ears gently. Fasten the throatlatch and noseband and, finally, lift the forelock over the browband and make sure the mane hair is not tangled under the headpiece. Check that all the straps are in their keepers and that the bit and noseband are at equal heights on both sides.

To remove a bridle, fasten a headcollar round the horse's neck and tie him up. Undo the noseband and throatlatch, then take the reins over his head. Holding the headpiece, slip the bridle gently over the horse's ears and allow him to drop the bit – being careful not to pull it out.

It's customary to tack up and untack from the near (left) side and most horses are used to this. It's logical to put on a bridle this way so you do not have to move from one side to another to fasten it, but when you get to know your horse and you are confident with each other, it's a good idea to accustom him to being saddled and unsaddled from either side so that you can work in a confined space such as a horsebox if necessary.

If you use a numnah and it is not already secured to the girth straps, put this on the horse's back slightly in front of where you want it to end up and place the saddle on top, making sure that the stirrup irons are run up to the top of the leathers so they do not bang against the horse's sides; the leathers should be tucked behind the irons so the latter do not slip down. Pull the numnah up into the gullet and slide it and the saddle back to the correct position, which will ensure that the coat hairs lie flat.

When a martingale or breastplate is used, it needs to be put in place now so that you can pass the girth through its attachment loop before girthing up. Double check that the numnah is pulled up and that the saddle is sitting centrally. If you are working from the nearside, it is better to go round to the off (right) side to let down the girth rather than letting it drop over, which may startle some horses and also means there is more risk that you will not notice that it is twisted.

Pass the girth through the martingale/breastplate attachment loops if necessary. Fasten the girth loosely to start with, then tighten it gradually until it is secure enough for you to mount without the saddle slipping. You should still be able to slip your fingers between the girth and the horse's side.

Before you get on, check the girth and if necessary tighten it again. Pick up each foreleg in turn and gently pull it forward; this will ensure that the skin and hair in the girth and elbow area lie smooth and there are no wrinkles that will cause pinching or rubbing.

Whenever possible, use a mounting block rather than mounting from the ground. This minimizes the chance of you pulling the saddle to one side as you get on, which will be uncomfortable for the horse, put you out of balance and, eventually, twist the saddle tree.

Check your girth as soon as you are settled in the saddle and again ten minutes after you have started riding. Some horses blow themselves out when the girth is done up and you may find that you are less secure than you imagined!

To remove a saddle, first run the irons up the leathers and secure them as before. Unfasten the girth on the nearside and go round to the offside so you can place the girth over the seat of the saddle – the side that goes next to the horse should rest on the saddle seat so that the leather does not get scratched or muddied by any dirt on the outside of the girth.

Lift off the saddle and numnah, if used; slide them back a little as you do so to keep the coat hairs lying flat.

Never leave tack where it can be knocked or chewed. If you carry it in your car, use a saddle cover for protection and rest the saddle with the seat upwards to avoid scratching the pommel or cantle. Some car manufacturers sell special saddle-carrying brackets that can be mounted in the back and folded up when not in use.

Protective boots

Boots designed to protect the horse's legs are a sensible precaution when doing fast work or jumping. It is also a good idea to use them when riding a horse who doesn't move perfectly (see Chapter 11) and may sometimes knock one leg against the other, or on a young or unschooled horse who is unbalanced, and when lungeing a horse. This chapter deals with boots used when a horse is being worked; Chapter 14 explains how to protect your horse's legs when transporting him in a horsebox or trailer.

Unless there is a specific reason for using boots, as explained above, you don't need to use them all the time. For instance, some people use them when they turn horses out in the field, but if worn for long periods they can cause more problems than they prevent, as a build-up of sweat or trapped dirt underneath them may cause skin problems.

The three most commonly used types of boots are brushing boots, tendon boots and overreach boots. Most are now made from high-tech synthetic materials that are easy to wash, but leather boots that need to be cared for in the same way as leather tack are also available. They come in various sizes to suit different types of leg conformation.

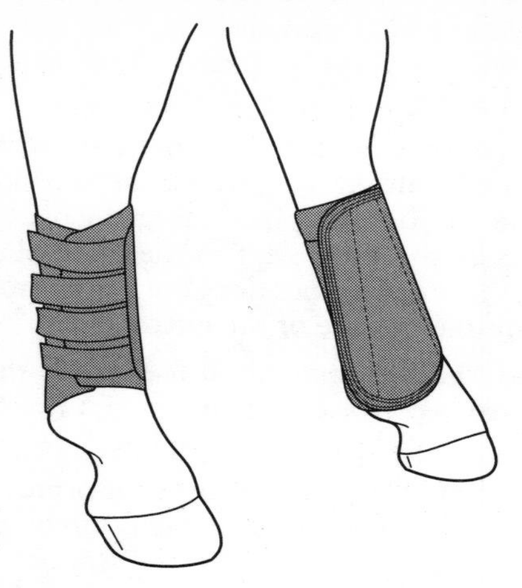

figure 9 brushing boots

Brushing boots fit on the lower part of the leg to protect the cannon bone area. They have straps which can fasten with Velcro or similar hook-and-eye materials, or buckles. They should be fastened snugly enough not to slip, but not so tight that they cause too much pressure.

These boots usually have three or four straps on those for the front legs and four or five on the hindleg ones. When you put them on, place them in position and fasten the centre strap first, as this prevents it slipping down the leg whilst you fasten the other ones. For the same reason, take them off by unfastening the top and bottom straps and leaving the third one until last.

Tendon boots are fitted to the front legs and, as the name suggests, are reinforced down the back to protect the tendon. They are often used when jumping and should be fitted with equal care.

Overreach boots are supposed to protect the horse from injuring himself if he overreaches – hits the heel of his front leg with the toe of his hind shoe. The problem with them is that if they are long enough to be effective, they are often also long enough to cause the horse to trip if he stands on the bottom edge. In most cases, it is best not to use them unless advised to do so by an experienced trainer or instructor.

Tack care

Looking after tack isn't just a matter of protecting your investment – it helps keep you and your horse safe. Weak stitching on a stirrup leather that goes unnoticed can lead to it giving way and causing a fall, whilst dirty tack can cause skin problems.

In a perfect world, all tack should be taken apart and cleaned thoroughly every time it is used and in professional yards with plenty of staff, this may be feasible. But for the average working owner, it may be impossible; a more realistic regime is to rinse the bit every time it is used, to avoid a build-up of dried saliva that may rub the corners of the mouth, and to clean off any obvious dirt, sweat and mud from leather or synthetic materials. You also need to make sure that your girth, numnah and protective boots are clean, so try and have enough so that you always have spares to use whilst others are being washed.

Once a week, back up your daily routine with a thorough clean. Take your bridle apart by undoing all the buckles, which will leave you with a headpiece, cheekpieces, noseband and reins and take the stirrups, girth and numnah from your saddle.

To clean leather tack, you need a bucket of clean, hand-hot water; two sponges and a cotton facecloth and a bar of glycerine saddle soap. Once a month, but not more often than that, use a good quality leather balm.

Use as little water as possible on leather, as if it gets soaked, it can stretch and become brittle. If your tack gets soaked from riding in the rain, let it dry out slowly in a room that is warm but not overheated; don't put it in front of a direct heat source such as a radiator or it will dry out too quickly and the leather may crack.

Use a sponge that has been dipped in clean water and wrung out to wipe off dirt or mud from the leather, rinsing and wringing it out frequently. Deposits of grease from the horse's coat are easier to remove with a facecloth that has been soaked, then wrung out – a more convenient method than the traditional way of pulling hairs from the horse's tail to make a cleaning pad!

When you've removed grease and dirt from both sides of the leather – remembering all the hidden bits such as the girth straps and undersides of the saddle flaps – use a second sponge to apply saddle soap. Old-fashioned grooms spat on their bars of soap, but it's probably more politically correct to put literally a couple of drops of water on it. The idea is to use as little water as possible, so it's better to wet the soap than the sponge; if the soap foams when you rub it into the leather, you've used too much water.

Rub the soap into both sides of the leather, particularly the rough, 'flesh' side. When you've finished, wipe over with a clean, dry cloth if necessary. Once a month, or at intervals recommended by the product manufacturer, use a leather balm instead of saddle soap.

Bits and stirrup irons can be washed in plain water and polished dry with a soft cloth. If you use rubber stirrup treads, the easiest way to remove trapped dirt is to scrub them with an old nailbrush. Boots, numnahs and synthetic girths should be washed according to the manufacturer's instructions.

Synthetic saddles should be cleaned as the manufacturer directs. Synthetic bridles are usually either wipe clean, machine washable or both.

Safety checks

Cleaning tack is a good opportunity to make routine safety checks, though it should be second nature to cast your eye over it for obvious problems every time you tack up. There are two potential weak areas: stitching and any place where metal rests on leather – in particular, the parts of the cheekpieces and reins which come into contact with the bit and the area of each stirrup leather which goes through the hole at the top of the stirrup iron.

Check stitching by getting hold of stirrup leathers in both hands and giving a good tug. If it is already coming undone, or starts to become undone when you do this, it definitely won't stand up to bearing a rider's weight. Another area to check is the stitching at the top of the girth straps.

Security and storage

Tack theft is big business, so as well as making sure that it is stored safely, you need to give it identifying marks; in the UK, postcodes are the best option, as stolen property which is recovered can be traced back to the owner.

Saddles are usually stamped on the flaps with a serial number, which should be recorded. It is also a good idea to buy an engraving tool – available cheaply at most DIY stores – and engrave your postcode on any accessible metal areas such as the stirrup bars at the top of the saddle, which take the stirrup leathers; bit rings; stirrup irons and so on.

Leather quickly goes mouldy if kept in damp conditions, so if you don't have access to a safe, dry tackroom, it's best to keep it in your home, out of sight. This may mean you end up with a spare bedroom that looks like a tackroom, but it's better than having tack that deteriorates or is stolen.

In this chapter you will learn:

- **what are the different types of rug**
- **how to fit a rug**
- **how to put on and take off a rug safely**
- **how to care for rugs.**

plate 1 ragwort in the rosette stage

plate 2 ragwort in the flowering stage

plate 3 Skewbald

plate 4 Cob

plate 5 Arab

plate 6 Thoroughbred racehorse

plate 7 three-quarter Thoroughbred

plate 8 Connemara pony

plate 9 Exmoor ponies in their natural habitat

plate 10 Highland pony

plate 11 Welsh Cob

plate 12 freezemark

plate 13 horse in a very poor condition

plate 14 horse in a good condition

There are so many types of rug available now that today's horse can easily have a wardrobe larger than his owner's. Fortunately, you don't have to buy everything – and certainly not all at once. In a few cases, such as ponies who grow thick coats and live out all the time in fields with good shelter, rugs can be kept to the minimum. Most horses, though, will need a variety throughout the year.

The main types of rug are turnout rugs, which are sometimes still referred to as New Zealands after the original canvas outdoor rug developed in that country, and stable rugs. You may also want or need to buy a thermal rug that can be put on a wet or sweating horse to transfer moisture from his body to the outside; a fly sheet for the summer to protect your horse from biting insects when he's in the field; a thin cotton summer rug, often referred to as a summer sheet and used for purposes ranging from keeping a horse clean to protecting him from draughts and flies in the stable and when travelling; and an exercise sheet, which covers the back half of a horse and can be used when riding in cold weather.

Turnout rugs are usually the most important on the list. Modern ones are made from high-tech waterproof materials that are also lightweight and breathable. They come in different weights, depending on how much warmth is needed, and it's essential to have at least two so that if one gets damaged, your horse has another to wear whilst it is being repaired.

Although some owners still like to have separate stable rugs, which are only worn indoors, others prefer to use lightweight turnout rugs indoors and out. It's a practical way of getting use from two turnout rugs and they tend to be tougher and more resistant to dirt than the fabric used for stable rugs, which are obviously not waterproof. If you follow this system, try and keep one rug for indoor use only, as otherwise dried mud will shed dust in the stable.

Thermal rugs which wick the moisture through to the outer layer of fabric are expensive, but once you have bought one, you will never want to be without one. They can be used for drying off, for travelling, when conditions are appropriate, or for adding extra warmth underneath another rug.

Summer sheets are useful for travelling horses in warmer conditions. They can also help you cut down on the number of heavier rugs needing to be washed, as if you use a cotton sheet next to your horse's skin you can wash this regularly in an ordinary washing machine, keeping the lining of the top rug clean.

Fly rugs are made from strong, lightweight mesh fabrics and help to keep flies and other biting insects from making your horse's life a misery. Ordinary fly rugs will usually not protect horses who suffer from sweet itch (see Chapter 09) though there are special designs which help. Many fly rugs come with neck covers to cover as much of the horse as possible. Turnout rugs and stable rugs are also available with neck covers, either built-in or as separate attachments that fasten to the main rug. They are useful for keeping the horse warm and for saving time on grooming, as they eliminate the risk of muddy manes. Be careful the design allows for your horse to stretch down his head and neck without putting pressure on the base of the mane; this is a particular problem with horses who have pronounced withers – in general, separate neck covers seem less likely to cause it than ones which are built in.

Exercise rugs fit behind and round or under the saddle and can be used to keep a horse warm and dry when you are riding out in bad weather. When you are riding on the roads in gloomy or rainy weather, an exercise sheet made from reflective, fluorescent material is a great way of making sure that other road users can see you from a long way off. It is, of course, always sensible for riders to wear high visibility gear in all weathers when riding on the roads.

Size and adjustments

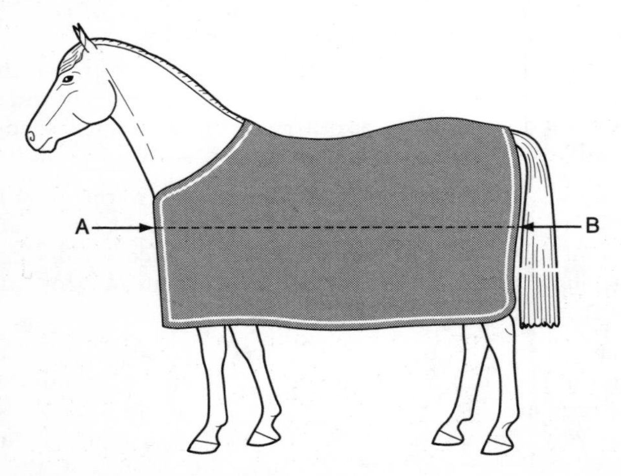

figure 10 how a rug is measured

For a rug to do its job, it has to fit properly. This depends on it being the right size for the wearer, well designed and correctly adjusted. Rugs are still measured in feet and inches, and the measurement is taken from the centre of the horse's chest, along the side of the body until you reach the point where a perpendicular line dropped from the top of his tail would meet it.

The one exception is the exercise rug, because it only covers the back half of the horse. This type is usually marketed in small, medium or large sizes – not always helpful, but the retailer should have a good idea of what size should suit your horse on the basis of his height, build and normal rug size.

Whatever type of rug you buy, try it on first over a clean cotton summer sheet or an old, clean cotton bed sheet spread over your horse's back and withers. This will keep it clean so that if you have bought the wrong size, the shop should be willing to exchange it.

When it is correctly adjusted, a rug should come just in front of the withers and stop at the top of the tail – though if there is a tail flap for added protection against the weather, this will obviously extend farther. It should come far enough down the sides so that you can't see the horse's belly. A rug that is too big will slip back and cause rubs, whilst one that is too small will restrict the horse's movement and, again, rub bare patches in the coat. Either will cause pressure points.

Design has nothing to do with colours and patterns, though manufacturers bring out new ones each season just as in the fashion industry! What does matter is that the rug is made to follow the contours of the horse's body but also to allow freedom of movement.

There are several different fastening systems for keeping rugs in place. All rugs fasten at the chest, either with buckles, metal T-bars or 'click' fastenings made from strong plastic, where one side clicks into the other, or rings which are clipped together by a harness system passing under the belly. If there are two fastenings rather than one, it allows for more leeway and there is less likelihood that the neck will gape.

The body of the rug is often held in place by cross surcingles, straps sewn to the sides which cross underneath the horse's belly and fasten diagonally on the other side. Alternatively, some designs have a system of straps, sometimes called spider harnesses, radiating from a central ring that fasten at the chest, go between the front legs and under the belly and are secured at the back of the rug on each side. Legstraps, which fasten from back to front round the tops of the front and/or hindlegs, are not as popular as they used to be, but are still found on some designs, particularly outdoor rugs.

Once you have made sure that a rug is the right size, it can be adjusted correctly by following these guidelines:

- When the chest fastenings are secured, you should be able to fit a hand's width between the rug and the horse's chest.
- Cross surcingles should cross in the centre of the horse's belly and you should be able to fit a hand's width between each surcingle and the horse.
- Legstraps should be fastened according to the rug manufacturer's instructions. The usual guidelines are that front legstraps should not be linked, but hind ones should.
- Rugs with harness fastenings should also be adjusted according to the manufacturer's instructions, as methods can vary.
- Check all fastenings daily to make sure the adjustments are still correct, as sliding fastenings such as those on cross surcingles gradually work loose.

On and off

To put on a rug with cross surcingles, first make sure that chest fastenings and cross surcingles, if fitted, are undone. Harness fastenings should be removed altogether and if the rug has legstraps, these should be clipped to the fastenings on the same side so they do not flap about when you put on the rug.

Next, fold it in half by taking the tail end to the front, so the lining is uppermost. Place the folded rug on the horse so that the front edge is in front of the withers. Now fold it back to the quarters, easing it back until it is in the correct position; putting it on slightly too forward and easing it back ensures that the coat hairs lie flat underneath it.

Once it is in position, fasten the cross surcingles first. This is a safety measure: if the horse spooks, the rug will not slip round and tangle under his feet, as could happen if you fastened the chest straps first. With the cross surcingles secured, fasten the front straps without dragging the rug forward and altering the lie of the coat.

You can now fasten the legstraps, if these are fitted. When fastening rear legstraps, stand behind but to one side rather than directly behind the horse, so if by chance anything startles him and he kicks out, you are not in firing range. If the rug has a harness fastening, place it on the horse as outlined above and fasten the harness as the manufacturer suggests.

To take off a rug, reverse the procedure. First unfasten any legstraps, taking them away from the legs and clipping each one to the side it originates from. Next, undo the chest fastening, followed by the cross surcingles. Finally, fold the rug in half from front to tail and slide it gently off the horse in a smooth backwards and sideways motion.

If the rug has an underbelly harness, follow the manufacturer's instructions. In most cases, you will need to unfasten it completely before folding the rug and sliding it off as above.

Cleaning, storage and security

The good thing about modern rugs is that most of them are washable. The less bulky ones will often fit in a domestic washing machine, but if the rug is fitted with metal fastenings it is a good idea to protect the machine from damage by tying old tea towels or something similar round them.

With all washable horse equipment, it's better to use non-biological washing powder or pure soap flakes. Biological washing powder may cause skin reactions in some animals – including people!

With rugs that are too bulky or simply too big to fit in your washing machine, you can either use a commercial rug washing service or invest in a power washer. Many saddleries now offer a rug washing service and there are also companies who will collect your rugs, wash them and deliver them back to you.

Small power washers intended for jobs such as washing cars can also make a good job of washing rugs, as the water comes out at high pressure. They are relatively inexpensive and will soon pay for themselves.

When you need to put rugs into storage, for instance after a winter's use, clean them first and store them in dry, vermin-proof containers. It is very annoying – and expensive – to find that your expensive New Zealand has been chewed by a mouse or rat!

Marking rugs with your postcode, written in large letters and numbers, acts as a deterrent against theft. There have been many cases of thieves stealing expensive rugs literally off horses' backs as they are grazing.

09 the healthy horse

In this chapter you will learn:

- **how to recognize possible problems**
- **how to monitor vital signs**
- **when to call a vet**
- **how to treat minor conditions.**

Keeping a horse healthy is a mixture of observation, correct management and preventive care. You need to be able to recognize signs of possible problems, but at the same time you need to get to know your horse so you know his physical characteristics and also when his behaviour or demeanour indicates that something might not be quite right. For instance, if you already know that he has a small splint (a bony enlargement on the inside of the cannon bone) you won't panic if you suddenly notice it.

Vets often call this the ABC of what to look for: appearance, behaviour and condition. There are also some useful guidelines that can help pinpoint the difference between a healthy horse and one who might have a problem. They include temperature, pulse and respiration, commonly known as TPR, and it's important to know not only how to measure them, but also what the normal values are for your horse so you can spot anything out of the ordinary.

Temperature

The normal temperature for an adult horse is 37–8° C. It is far easier to take a horse's temperature with a digital thermometer than with an old fashioned 'shake down' one, so buy one for your equine first aid kit (see later in this chapter). You can get special ones for veterinary use but those sold for people are fine. As you take a horse's rectal temperature, it's important to keep his thermometer separate from your household one!

Try and get someone to hold your horse and reassure him whilst you take his temperature; if this is impossible, tie him up. Lubricate the insertion end of the thermometer with petroleum jelly, then stand to the side and rear of the horse – not directly behind him – and run your hand down his quarters so further contact does not take him by surprise.

Insert the thermometer gently into the rectum and tilt it slightly to the side, then wait for the signal that the reading is completed. Once you have taken the reading, clean the thermometer with a solution of disinfectant diluted in cold water.

It's a good idea to take your horse's temperature regularly, at a time when he is relaxed and in good health; don't take it after he has worked, as it will naturally be slightly higher after exertion. This will give you a guide to your horse's normal temperature and enable you to spot a slight rise that may be one of the signs of illness.

If you suspect that he is off colour or ill, it is always a good idea to take his temperature. If the reading is 40° C or more, get veterinary advice.

Pulse

The normal pulse rate for an adult horse at rest is between 35 and 42 beats per minute. The easiest way to measure this is to take the resting heart rate with a stethoscope on the left-hand side of the lower chest, near the elbow; an alternative method is to place your first three fingers under the jaw, behind the lower edge, until you feel the facial artery (this is often difficult to find). Rather than take the pulse for a whole minute, count the beats in 15 seconds and multiply by four.

It is also useful to know where to feel for the digital pulse. This is often hard to find in a healthy horse, but a strong digital pulse is often associated with inflammation of the foot, as in laminitis. The easiest way to locate it is to apply light pressure with two fingers at the side of the pastern, just below the fetlock. If you can't find it, ask your farrier or vet to show you.

Respiration

A horse at rest will normally take between 8 and 20 breaths per minute. The easiest way to count the respiratory rate is to stand behind and to one side of the horse and count the number of times his flank moves. Count each inhalation and exhalation as one breath.

There are many other indications that give clues to a horse's state of health, but they need to be tempered with common sense. For instance, whilst it is often said that a dull, staring coat is a warning sign – and it can be – don't confuse this with the thick winter coat of an unrugged native pony, as hairs standing on end to trap a layer of air is nature's way of providing insulation.

However, things you need to watch out for include the following.

- If the horse is sweating and there is no obvious reason, he may be in pain. Sweating is sometimes associated with colic (abdominal pain) which is always serious.

- The horse who is normally interested in his surroundings and suddenly becomes dull or lethargic may be ill.
- Discharge from the eyes or nostrils should not be ignored.
- The mucous membranes on the gums and around the eyes should be a salmon pink colour, though some horses' gums naturally have dark pigmentation. This is one of the scenarios when knowing what is normal for your horse will warn you when something is wrong.
- Most horses normally have a healthy appetite. If yours suddenly goes off his feed, something is wrong.
- Keep an eye on the quantity and consistency of your horse's droppings. If you suddenly find that your horse is passing fewer droppings than normal, and/or they become hard and impacted, he may be constipated and/or showing early signs of colic. When a horse is excited, or when the spring grass comes through, his droppings will often become looser, but if this continues, it should be regarded as a warning sign that all may not be well.
- Make sure your horse is drinking regularly. If his water intake suddenly decreases, he will become dehydrated, which can lead to serious problems.
- The normal colour of a horse's urine ranges from pale to brownish yellow. If this suddenly changes – and especially if the urine takes on a red tinge, indicating the presence of blood – something could be wrong. You should also get veterinary advice if your horse is trying to urinate but having difficulty.

Routine care

Protecting your horse against disease through vaccination, regular dental care and good farriery is essential to keep him healthy. This applies whether he is a horse in work, a youngster who is not old enough to be ridden or a retired horse.

Vaccination

It is vital that all horses and ponies are vaccinated against tetanus and equine influenza. Some people believe that horses who are not working and do not leave the owner's premises, such as retired horses, do not need protection against equine flu, but vets say they are still at risk if they come into contact with horses who have been exposed to flu viruses, even if they are vaccinated.

Tetanus is a killer; tetanus bacteria are prevalent in the soil and can enter the horse's system through the tiniest nick or graze. Because it is a bacterial infection, not a viral one, it can't be passed from one horse or species to another. However, humans can also contract tetanus, so it is important that you are also protected against it.

Equine flu is a viral infection that – unlike bird flu – can't be passed to other species, but is a serious illness that can lead to permanent damage to the horse's respiratory system and even death. As a simple and relatively inexpensive programme of routine vaccinations can prevent both, there is no justification in not vaccinating your horse. If you want to compete, you will also find that many organizing bodies insist that all animals are vaccinated.

Vaccinations can only be given by vets and have to be administered within certain time limits. Your vet will enter the details in your horse's passport and, if you miss the time limit, you will have to start a new programme to fulfill competition rules.

There are two sets of requirements, one set by the Jockey Club and the other by the FEI (Federation Equestre Internationale), which is the international body in charge of horse sports. The only difference is the timing and frequency of vaccinations; discuss the issue with your vet, but you will probably find that unless you want to compete at top level where international competitors will be present, Jockey Club requirements will be fine.

Jockey Club rules stipulate that the first two vaccinations in a primary course must be given at an interval of between 21 and 92 days and the third one 150–215 days after the second. When this is completed, the horse will need an annual booster, the first to be given within one calendar year of the date of the third vaccination.

FEI rules follow the same time line for the first two vaccinations, but the third one and subsequent boosters must be given within intervals of six calendar months.

Other vaccinations are available, including ones against strangles (see later in this chapter) and equine herpes. Your vet will advise you on whether your horse needs extra protection, as there are several considerations to take into account.

Dental care

Horses' teeth differ from ours in that they grow throughout their life. Although they are worn down by eating, the diet of the domestic horse is different from that of the wild one and nature needs a helping hand.

Adult male horses usually have 40 teeth and mares have 36. All have six front teeth, or incisors, in each jaw; 12 cheek teeth, or premolars and 12 molars. Males also have four tushes, one at each side of the upper and lower jaw, which are what remains of canine teeth. Occasionally, these are also found in mares, but are usually smaller.

A horse's lower jaw is narrower than his upper one, so when he grazes, the outside edges of the upper cheek teeth and the inside edges of the lower ones may be worn in a pattern that causes sharp edges that can graze the cheeks or tongue. The cheek teeth in both jaws can also be worn into a pattern that results in sharp hooks. These interfere with his ability to chew, so not only does he not get maximum benefit from his food, he may also develop digestive upsets or colic. Sharp teeth also interfere with the action of the bit, which will make the horse uncomfortable and resistant.

Horses of both sexes often develop shallow rooted, vestigial premolars called wolf teeth. These too can interfere with the action of the bit and are often removed as standard to prevent this happening – it is a relatively minor procedure when done by a vet or qualified equine dental technician, as the shallow roots mean they can be taken out easily.

In theory, anyone can set up as an equine dentist – though someone who caused harm would be liable to prosecution – so it is important that you use a qualified person. All vets can carry out dental work and some have a special interest in it; others prefer to recommend their clients to qualified practitioners who are not vets. In the UK, there are equine dental technicians registered with the British Equine Veterinary Association. (See Taking it further for contact details.) If in doubt, always ask your vet for advice on organizing your horse's dental care.

Opinions vary on how often dental checks should be carried out. The sensible approach is to call in your vet or equine dentist whenever you suspect a problem and take guidance on intervals between check-ups. Warning signs include a horse who tilts his head to one side when ridden, starts resisting the bit or drops food out of his mouth when eating.

Worming

All animals harbour parasites in their bodies. Horses are particularly at risk from intestinal worms, so routine worming measures are also essential. To be accurate, they should really be called anti-worming measures, but as the drugs used are generally referred to as wormers, most people use the first term.

One of the best ways to protect your horse is to remove droppings from the field as often as possible, ideally every day. This, combined with a correct worming programme, should keep him in good health. Don't assume that because a horse looks healthy, he does not have a heavy worm burden.

Problems caused by parasites range from weight loss and diarrhoea to severe colic. A foal will not have worms when it is born, but as soon as it starts grazing, it becomes susceptible. Worms are parasites and, as such, they depend on a host to maintain their life cycle.

Many drug companies suggest worming plans, but although these can be useful guidelines, the best way to check that you are giving your horse the best protection without wasting money or administering unnecessary medication is to talk to your vet. He or she can advise you on the best strategy for your circumstances – for instance, the owner of two horses, kept at home on well-managed grazing, can follow a different plan to someone whose horse is on a large yard with a changing equine population.

If you buy a new horse, it is best to ask your vet's advice about which drugs to use to start with, unless you can be absolutely positive about its worming history. There is a difference between worming regimes designed to treat a horse who has a worm burden and those designed to maintain health. Your vet may suggest taking a faecal sample and perhaps a blood sample to establish the horse's current state and will devise a worming strategy based on this.

Most worms which affect horses are species specific, which is one reason why it is a good idea to rotate grazing when possible so that it supports sheep or cattle instead for short periods. Although there are several types of worms, most have a similar life cycle. Microscopic eggs and larvae attach to the grass, are taken in when the horse grazes and develop into adults in the guts. The adults lay eggs, which are passed out, together with larvae, in the droppings.

The commonest worms in horses are small redworms (*Cyathastomes*). These are dangerous because they can hibernate in the gut wall and become encysted, their numbers reaching a peak in the late autumn. When they become active in the spring, huge numbers may erupt from and damage the gut wall.

Large redworms (*Strongylus vulgaris)* have a different life cycle but are equally harmful. Large roundworms or ascarids (*Parascaris equorum)* can reach up to 40 cm long. Foals to two-year-olds are particularly susceptible to them and as their journey through the body takes them through the lungs, they can cause coughing and other problems.

Lungworm (*Dictyocaulus arnfieldi)* used to be a real problem in horses that were grazed with donkeys, or on land where donkeys had been. Although they did not seem to cause many problems to the donkeys themselves, these worms lived in horses' lungs and caused bad coughs. Modern worming drugs are effective against them.

The tapeworm (*Anoplocephala perfoliata)* uses the horse as its second host. Its first is a tiny forage mite which is picked up through grazing and hay – so your horse is vulnerable all the year round. The adults shed rectangular segments which may be seen in the dung, although as with all worms, the fact that nothing is visible to the naked eye does not mean that your horse is not affected. Tapeworms are a major cause of some types of colic.

Pinworms (*Oxyuris equi)* are thought not to cause internal problems, but will make the horse itchy around the anal area. Eggs in a sticky substance are laid round the anus and if seen should be wiped off with a damp cloth.

The other common parasites are bots (*Gasterophilus intestinalis)* which are not worms but the larvae of the bot fly. The fly lays sticky, clearly visible eggs on the horse's coat; the horse then licks these off and so transfers the eggs to its digestive system.

If you are on a livery yard, the only effective system is for the yard owner to get veterinary advice on working out a suitable programme and insist that every client follows it, so that all horses are wormed on the same day. Worming measures only work if all the horses grazed in the same field are treated in the same way, though new arrivals' state of health should first be assessed as outlined earlier.

If you kill the four main groups of parasites – roundworms, redworms, tapeworms and bot fly larvae – you will also take care of any others that your horse may be harbouring. This means using the appropriate drugs at the appropriate time. At one time, it was necessary to use several different types of drug to target different types of parasite, but now there are wormers that are effective against most of them once you are able to put your horse on to a standard regime. There are different opinions on whether or not horses can build up resistance to them, but your vet will be able to advise you on the latest evidence.

Wormers are administered orally, either in the horse's feed or syringed directly into the mouth. Some companies now make them in palatable flavours such as apple or caramel.

Farriery

A good farrier plays an essential role in keeping your horse sound. Even horses who are not shod need to have their feet trimmed every six weeks, as the hoof grows at an average rate of 12 mm per month. Growth rate varies according the time of year, the horse's diet and other factors.

Some people believe that all horses can work without shoes, but this view is still a minority one. Rather than be swept along by evangelical statements, talk to your farrier and vet about your horse, his lifestyle and the conditions in which he is worked. Horses who are hacked regularly on roads and hard tracks usually need shoes, though there are undoubtedly exceptions.

In the UK, all farriers have to be licensed with the Farriers Registration Council and only a registered farrier can shoe a horse or trim a foot if the trimming is carried out in preparation for a shoe being put on. This means that owners can legally trim their own horses' feet if they are not intending them to wear shoes; some do, but it is a controversial area and it is recommended that you only entrust your horse's hoof maintenance to a registered farrier.

Hoof trimming is not as simple as it may look, as there is much more to it than taking off excess growth. The feet should also be trimmed in a way that maintains correct angles between hoof and limb, or strain will be put on the limbs that could result in lameness and permanent damage.

If you keep your horse on a livery yard, you will find that one or more farriers visit regularly and it should be easy to find one prepared to take on another client; if you have any problems, the FRC lists farriers in different areas (see Taking it further). Most other countries do not have regulations governing farriery, though many are investigating bringing in ones similar to those in the UK.

Owners also have a responsibility to pick out their horses' feet at least once a day, preferably twice. This means you clear out the muck and dirt that impacts in the foot and which can cause a foot infection called thrush if ignored. At the same time, you should spot any objects lodged in the foot that could cause injury and check for signs of loose shoes and risen clenches.

Clenches are the securing 'hooks' formed when the nails have been driven in, the ends twisted off and the remainder of the nail shank hammered over. If they work up again, this leaves a sharp projection that the horse could injure himself on if he strikes the foot against another leg. It is usually possible to tap these down with a hammer, but ask your farrier to show you how to do it safely.

If you find a shoe has become slightly loose, ask your farrier to make an emergency visit to re-fit it. Don't ride the horse until this has been done, because if it comes partially off whilst you are riding he may trip and injure himself on the shoe or a nail.

Removing a shoe

It is a good idea to learn how to remove a shoe yourself in case of emergency, which means you need a farrier's buffer and pincers – tools you can buy through your farrier or from some tackshops. A buffer has a chisel-shaped blade used to cut off clenches and is easier and safer to use than an ordinary chisel; pincers are used to grasp and pull off the shoe.

Before you start, make sure you are wearing tough legwear, ideally denim jeans with suede chaps over them that cover the full length of the leg. This will give you some protection if the horse pulls back and you come into contact with the shoe. The basic technique for removing a shoe is described below, but you should ask your farrier to show you.

To remove a front shoe, you need to pick up and hold the foot safely. With a front foot, stand at the horse's shoulder and pick up the foot as if you were going to clean it out. You need to place the foot between your knees to remove the shoe easily and safely, so step forward on the leg closest to the horse, bring the hoof under your knee and hold it with both hands. Bend your knees and bring them together, turning in your toes so the foot is held firmly.

With a hind foot, again pick it up as if you were going to clean it out, then stand so you can place the horse's lower leg across your thigh nearest to the horse. Rest the foot on your opposite knee.

To remove the shoe, cut off the clenches cleanly with the buffer so that the foot does not get damaged as the nails are pulled back through. Then use the pincers to ease the shoe loose, first at the heels and then along each of the sides. When this has been done, grip the toe of the shoe in the pincers and pull off the shoe.

First aid box

Every horse owner should have two first aid boxes, one to keep on the yard and one to keep in the horsebox or towing vehicle when out and about. Don't forget to keep one for humans, too; keep them in different coloured containers so they don't get confused (remember that thermometer?) Tape a card, on which you have written your vet's telephone number, to the inside of the lid – this should also be programmed into your mobile phone and be on display somewhere anyone on the yard can find it. Ready-stocked first aid kits are available, but you need to keep the following supplies available:

- digital thermometer
- moist wound gel, which will protect and cleanse at the same time – the proper name is hydrogel but it is marketed under brand names such as Vetalintex® and Derma Gel®
- large roll of cotton wool
- antiseptic such as Pevadine® or Hibiscrub®
- scissors with curved ends
- antiseptic wipes
- gamgee (cotton wool sandwiched between layers of gauze)
- poultice, such as Animalintex®
- clean bowl or small bucket
- duct tape for securing bandages and poultices
- dressings – non-stick dressings to cover wounds; self-adhesive bandages; 'cool' bandages which can be soaked in cold water and keep their cooling action when applied to the leg; cotton stretch bandages.

When to call the vet

Deciding when you can deal with a problem yourself and when you need to call your vet isn't always easy, so if in doubt, call. You will never be accused of wasting someone's time, because it is better to be safe than sorry. A vet would rather be called to look at a wound that might need stitching and find that it doesn't than have to deal with a serious infected joint because the horse's owner ignored a small puncture wound.

Some situations always count as emergencies. Never hope that the following will go away, but get immediate veterinary help. Try not to panic, however difficult that might seem, because it won't help your horse and if you are agitated, he will pick up on it.

- Colic means abdominal pain, but can be a life-threatening condition. Signs may include the horse turning round to look at or nip his flanks, sweating, pawing and rolling. If your horse seems listless and has not passed any/as many droppings as usual in the past few hours, he may be suffering from colic related to a build-up of impacted faeces. If you think your horse has colic and it's safe to do so, take his temperature before calling your vet. If the horse is lying down and/or rolling and is in obvious distress, don't take risks. Ring your veterinary practice immediately – all operate emergency call-out services and if the horse is ill during the night, leaving it until the following morning might be too late.
- Small scrapes and cuts can usually be dealt with via basic first aid, but there are times when it is essential to call your vet. These include the following scenarios, but always remember that if you are not sure, it's better to play safe.
  - Any wound, however slight, on a horse who may not have had protection against tetanus.
  - Any wound that is bleeding heavily or may have affected an artery (when the blood will be bright red and will spurt rather than flow).
  - A cut that has gone right through the skin and is gaping. Even if it is too small to be stapled or stitched, the horse may need a course of antibiotics. This applies particularly if the injury may have occurred some time before it was spotted, for instance if a horse is injured overnight and the wound only spotted when you visit next morning.
  - Wounds in the area of a joint or tendon should not be left to chance, in case it has affected the structure.
  - Puncture wounds can cause far bigger problems than their size suggests, especially if dirt has worked its way in.

- Eye injuries should never be left to chance. If there is an obvious wound, or if your horse has a partly or completely closed eye, treat it as an emergency and send for your vet.
- If a horse is severely lame or seems unable to put his foot to the floor, call your vet. If he can't put his foot to the floor, don't try and force him to move. Instead, stay with him whilst you wait for the vet to arrive. At worst, it could be a limb fracture, but it could also be something such as an abscess in the foot, so try not to panic.
- Laminitis is a painful foot condition associated with causes ranging from obesity to concussion. Is often thought of as a disease affecting little fat ponies, but can affect all types of horses. Signs of a severe occurrence include a reluctance to move and even to take weight on the front feet and a strong digital pulse. Early signs may include a shorter stride and a stronger digital pulse. If you even suspect laminitis, call your vet – if left, the condition will worsen and structures inside the foot may collapse. The worse scenario is that a horse will have to be put down for humane reasons. Treatment may include remedial farriery and your vet will advise you how to adjust your horse's diet and management.
- Occasionally, a horse may get food stuck in his gullet. This is known as choke, for obvious reasons – he will cough and heave and you will probably see food and saliva coming from his nostrils as well as from his mouth. Take him away from all sources of food and water and call your vet as a precaution; usually the blockage clears by itself, but you can't take chances.
- If a horse is showing signs of shock, perhaps after being involved in an accident, call your vet to check him over. Signs of shock can include a temperature that is below normal and gums that are much darker than their normal pink colour, as well as the horse's demeanour.
- Equine rhabdomyolysis syndrome, or ERS, is a long name for what is colloquially called azoturia, or tying-up. In simple terms, it is severe cramping of the muscles and may happen after the horse finishes work, or whilst you are riding. In mild cases, the horse will be stiff and take short steps behind; in severe ones, he may even collapse. Put a rug or, if you are away from home, coats over his back and hindquarters to keep the muscles warm. If he can't move or collapses, call your vet for emergency help. In milder cases, you will be able to get him back to the yard, either leading him or calling for transport home. Call your vet even if he recovers

straightaway, as the condition often reoccurs and possible causes need to be investigated.

- Strangles is a highly infectious bacterial throat infection. Its name comes from the fact that, in severe cases, the throat glands swell and pus-filled abscesses may appear in glands on the head. It is not a sign that an owner or yard operates poor standards of care or hygiene, and symptoms may include a high temperature, nasal discharge and swollen glands. Most horses recover quickly and without lasting problems. Unfortunately, some horses may be long-term carriers of strangles without showing signs of illness, which makes it easy for the disease to spread. Whilst strangles may not be an emergency in terms of the threat it poses to most horses' long-term health, the fact that it spreads so quickly means you should call your vet at the slightest suspicion so that quarantine measures can be taken. No horses should go off the yard or be allowed on it until your vet gives the go-ahead and everyone dealing with the horses should follow strict hygiene procedures – again, your vet will explain these.

If your horse has an injury or illness that needs veterinary attention, your priorities are to make sure that he is in safe surroundings and to call the vet immediately. Hopefully, you will be able to talk to the vet who will attend, or another vet at the practice, and will be given instructions on what to do until help arrives.

But although horses may seem to be accidents on four legs waiting to happen, and the list of possible emergencies makes horrifying reading, dangerous scenarios are fortunately rare. Most mishaps can be dealt with through basic first aid and preventive measures, and learning how to deal with simple wounds, to identify lameness and to spot the first signs of common problems will help prevent minor accidents turning into major ones. It will also mean that you can spot potentially dangerous conditions, or those that are not necessarily dangerous but could be transmitted to others, at an early stage.

Wounds

To clean a graze or small cut, hose it gently with cold water. Some people prefer to use a dilute antiseptic to clean wounds, but it must be diluted at the recommended strength – if it is too strong, it could damage the tissues and impede healing. If the wound is as insignificant as you thought, apply wound gel to protect it and, if you can do it safely, clip the hair away from the edges with a pair of round-ended scissors to confirm this.

Wounds on the lower limbs are more likely to get dirty, so it is best to cover them. Place a sterile non-adhesive dressing over the wound, then cotton wool or Gamgee®, followed by a stretch bandage and a cohesive bandage to keep everything in place. Wounds in other places are more difficult to bandage, but keep them as clean as possible.

There may be times when you need to poultice the sole of a foot to draw out infection. This is best done using a proprietary poultice such as Animalintex®, which should be prepared according to the manufacturer's instructions but must not be too hot.

Protect the horse's heels from getting rubbed by smearing them with petroleum jelly, then put the poultice in place. Place a layer of Gamgee® over it and hold it in place with a cohesive bandage applied in a criss-cross pattern – it's best to use the whole bandage for security. Finally, add either a top layer of criss-crossed duct tape or a commercial poultice boot.

Change the poultice every 24 hours and don't keep a foot poulticed for longer than three days, or you will soften the foot too much. You will need to keep the horse stabled to keep the poultice in place, though it is possible to buy 'hoof boots' that are said to be suitable for outdoor use over poultices.

Identifying lameness

Although it is relatively easy to spot that a horse is lame, because his gait will not be normal, the fact that he has four legs means you may have to look carefully to work out which is the affected limb. If there is a wound, heat or swelling it's usually pretty obvious where the problem lies, but the majority of lameness problems are in the foot, with no outward sign. To complicate matters, a horse can also be lame on more than one limb at the same time.

Except in very severe cases, lameness does not show when the horse is walking, so it is necessary to trot him up on a headcollar to assess the problem. If he is unwilling or unable to put a foot to the ground, or is lame in walk, call your vet immediately and don't attempt to persuade the horse to trot.

Lameness in a front limb – a term which includes the leg and foot – is easiest to spot. Take the horse onto a safe, hard surface and watch him being trotted away and then back toward you. He should be led on a headcollar and loose lead rope, so the movement of his head is not restricted.

A horse who is lame on a front limb will lift his head as the lame leg hits the ground and nod as the sound one does the same. The easiest way to remember this is that his head will 'sink on the sound leg'.

If a horse is lame on both front legs, he may actually appear sound, but will move with a shorter stride than usual. It's said that wicked horse traders of bygone times who wanted to sell a lame horse would deliberately lame him on the other side, too, so unwary buyers would not spot the problem!

Hindlimb lameness isn't quite so easy to identify. You need to see the horse trotted away from you and watch the movement of his hips, as the hip on the lame side will rise and drop more obviously as the horse tries to minimize the amount of weight he has to take on it.

Conditions affecting the skin

There are several conditions which affect the skin and, in all cases, you need to call in your vet to make sure of an accurate diagnosis. The commonest are ringworm, mud fever and rain scald and sweet itch.

Ringworm

This is a fungal infection and nothing to with worms. It is highly infectious and although it does not seem to bother the horse, is extremely annoying for owners. Not only does it spread rapidly from horse to horse, it can also be passed from horse to human. It is never fatal and can't be called an emergency, but should not be ignored.

Early signs are usually raised tufts of hair with a slight swelling underneath that develops into flaking patches of skin. They are often found on areas which come into contact with tack, such as round the girth or on the head. It is spread by direct contact, not just from horse to horse but via tack, rugs, grooming kit and – because spores can remain dormant for a long time – even fencing or stables.

The incubation period is between one and four weeks and if it is untreated it will eventually clear up, but this is not a responsible option. Your vet will take a skin scraping to confirm ringworm and provide medication, topical solutions or both. You will need to disinfect everything that has been in contact with the horse with a solution proven to kill ringworm, which your vet will advise you on.

One of the best ways of preventing ringworm is to make sure that each horse has his own tack, rugs and grooming equipment and that they are not swapped between horses.

Mud fever and rain scald

These are both inflammation of the skin caused by a bacterium called *Dermatophilus congolensis,* found in the soil and on the skin. Normally it is harmless but, if the skin gets too soft through being constantly wet – as can easily happen in muddy fields – it can cause inflammation leading to sores and crusty patches. It may also gain entry through a graze or wound.

They are essentially the same condition, referred to as mud fever when the legs are affected and rainscald when inflammation affects the body. In either case, you need to ask your vet to look at your horse in case he needs antibiotics; you will also need to trim the hair from affected areas and wash them with antiseptic solution every day until the condition clears.

The horse will need to be kept in dry conditions until the skin heals, which usually means stabling him. Your vet will also advise you on preventive measures, which may include using a barrier cream.

Sweet itch

Horses with sweet itch suffer from severe itchiness, as the name suggests. They will often rub themselves raw, especially along the mane and at the top of the tail. Sweet itch is an allergic reaction to the saliva of a type of midge called *Culicoides* and though it is more of a problem in spring and summer, climate change means some horses are affected for much longer periods.

The best way to help susceptible horses is to stable them at dawn and dusk, when the midges are most active. Use insect repellent at all times – the most effective is usually one containing DEET – and if necessary protect your horse with a special rug that the midges cannot bite through.

Ordinary fly rugs are not necessarily a defence against *Culicoides*, even though they may provide relief from the unwelcome attentions of other insects; look for close-fitting rugs sold specifically for sweet itch sufferers. Choose one that covers as much of the horse as possible; midges will bite the belly and head as well as along the body, so you may need a rug that incorporates a belly wrap and a hood.

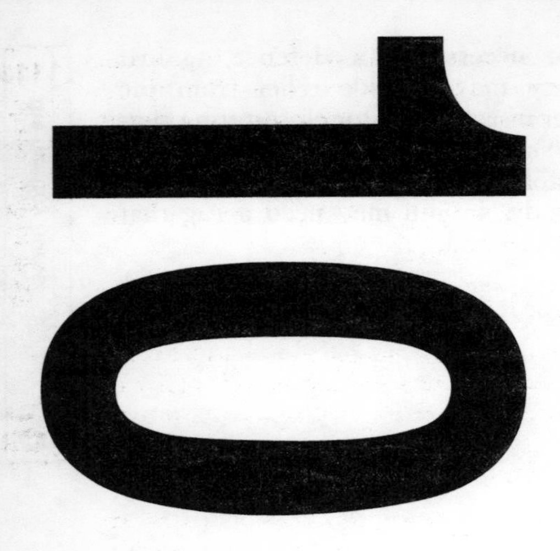

10 share, loan or buy?

In this chapter you will learn:

- **how to work out which option is best for you**
- **why formal share or loan agreements are important**
- **what are the different ways of buying a horse**
- **what legal protection may be offered.**

Once you have decided that you are ready to take on a horse and have made as many advance arrangements as possible, there is one final part of the equation to work out before you start reading the advertisements for likely sounding prospects. Are you going to start out by sharing someone's horse in return for sharing costs and work; are you going to take a horse on loan for an agreed period; or are you going to buy?

All three options have advantages and disadvantages. The obvious advantage of sharing or loaning a horse is that you don't have to make the financial outlay and get the chance to see if making a full-time commitment works out. However, you will have to accept that the horse's owner is in charge and you may feel that you would prefer to take on extra costs and responsibilities in return for the freedom this offers.

One arrangement that rarely works out, because it can lead to so many complications, is to buy a horse in partnership with a friend. No matter how well you get on and how much planning you do, there is so much that can go wrong. What happens if one partner becomes ill, has to move house or has a change in family and work commitments? How would you manage if one of you liked the horse and the other decided it wasn't suitable? These issues could, of course, be resolved on the basis of a pre-purchase written agreement – but it's usually better for one person rather than two to be the legal owner.

If you enter into either a share or a loan arrangement, it is important to have a written agreement. Some people prefer to have loan agreements, especially those intended to be long term, drawn up by a solicitor, feeling that whilst it means an additional cost, it protects everyone involved – including the horse.

It's vital that you feel happy handling and riding the horse, or, in the case of a children's pony, that rider and pony show the potential to establish a happy partnership. So whilst it may be tempting to compromise on some things because you aren't actually buying the animal, that can be a mistake.

When you are looking at a horse with a view to sharing or loaning him, ask yourself this question: if I was looking at a horse to buy, would this be one I would be keen to own? If the answer is no, the arrangement probably wouldn't work in any form. It's unfair to all concerned to think that you might as well give it a try, simply because you think you can always back out.

Sharing

If you have never owned a horse before, a share arrangement can be an ideal starting point as long as you find the right horse and the right owner. For instance, if you have been taking lessons at a local riding school you might find that someone who keeps a horse at livery there is looking for financial and perhaps practical help in return for an agreed amount of riding. In this situation, the yard owner should know all parties involved – including the horse – and be able to advise on whether it is worth pursuing the idea.

It is unfair to look on a share arrangement purely as a test run, unless the horse's owner is happy to do this in return for short-term help. Ideally, you should both be prepared to commit to a period of six months to a year.

Another possibility might be to lease a horse that you have been riding from your riding school, either full-time or for an agreed number of days per week. The advantages of this are that you have the security blanket of being familiar with the horse and having people around who know him, giving you the chance to get used to a degree of responsibility. The disadvantage is, again, that you will usually be limited in what you can do with him.

Before you enter into a lease or share arrangement, talk to the horse's owner about exactly what would be required of you, what you hope to do and whether this would be agreeable. You don't want to find three months down the line that you had hoped to be able to compete in some local competitions and the horse's owner is not prepared to let you.

Below is a list of some of the points to consider. If initial discussions go well, draw up an agreement, perhaps with the proviso that it could be cancelled by either party without notice during the first month if things don't work out. Even if this happens, expect to pay a pro rata share of any costs that have been agreed.

It is unwise to enter into any form of share, loan or leasing agreement unless the horse is insured for veterinary fees and for public liability insurance. If insurance is already in place, make sure it applies to you, as well as to the horse's owner.

Although targeted to a share agreement, the following questions to consider are also valid if you are thinking of leasing a horse part-time.

- What percentage of costs will you be expected to pay? These should be detailed and will usually include all routine maintenance, including livery, shoeing, insurance, worming and vaccination.
- Who pays for maintenance and cost of equipment, such as replacement tack and rug repairs? In most cases, it is sensible for these to be borne by the owner, as they will be his/her property.
- What happens if the horse needs veterinary treatment? The most workable and sensible arrangement is usually for the horse to be insured for veterinary fees and the excess of any treatment costs – the amount paid before the policy benefit kicks in – to be shared. If this is the case, make sure the excess is agreeable to both of you.
- What happens if the horse is off work because of illness or injury? One option is for the sharer to agree to pay costs for a certain time and to review the share agreement after this.
- If the share agreement covers looking after the horse, what duties are to be carried out by each person, and when? There has got to be a certain amount of flexibility here, but you don't want to find that you are the only person picking up piles in the field, or that the horse's owner is fed up because you never clean his tack.
- How often will you be able to ride the horse? Will he be available to you on the same days each week, or do you want a more flexible arrangement? The important thing is to make sure that ill-feeling won't arise because, for instance, the owner wants sole use of the horse at weekends and you would also like to ride at that time.
- What are you allowed to do – and, perhaps even more important, is there anything the owner does not want you to do? It would certainly be a good idea to have lessons from a good instructor; if the owner also has lessons, it would be best if you both went to the same person to ensure continuity. If the owner doesn't have lessons, you must make sure that your choice of instructor is acceptable. The downside is that, in some situations, you may feel you aren't making as much progress as you would like because the horse's owner isn't interested in following the same approach.

Loaning

Taking a horse on loan can, in some situations, be an excellent arrangement for all involved. For instance, you may be lucky enough to find someone with a much loved horse who they do not want to sell – usually because he is an older animal who has given a lot of pleasure and the owner wants to safeguard his future – but who wants him to keep active and get lots of attention. Understandably, this sort of loan agreement will often carry the stipulation that the horse must remain at his present yard so the owner can keep an eye on things.

If you are looking for a children's pony, you may also find one who has been outgrown and needs a temporary home before returning to his owners when a child who is currently too young to ride is ready to start, or retiring.

It isn't a good idea to think about looking for a horse on loan simply because you can't afford to buy one. If this is the case, it's highly likely that you will not be able to afford the 'running costs' either; wait until your finances improve, or think about a share arrangement where you only have to pay part of the costs.

If loaning seems a good option, you can either look for a horse offered on loan from a private home or investigate the loan schemes offered by reputable equine rehabilitation charities. In the UK, the best known are the International League for the Protection of Horses, which has four centres throughout the UK, and the Blue Cross, which has equine centres in Oxfordshire and Kent. Other charities include ones specializing in re-training and re-homing ex-racehorses, but such horses are usually suitable only for more experienced riders.

All charities operate stringent checks to ensure that horses are only entrusted to suitable borrowers; these include inspecting the premises where the horse will be kept and, if you are hoping to take on a horse for riding, assessing your capability. It may be a long time before a potential loan horse is available and it is in the horse's interest for its loan home to be as long-term as possible.

Horses come to centres such as these for a variety of reasons, but many will have had physical or behavioural problems. However, the skill of staff at centres such as the ILPH is such that many horses previously written off as no-hopers are rehabilitated and go on to successful careers, whether as pleasure horses or competition animals.

Even when a loan horse is kept at his present yard under his owner's management, you need to be clear about every aspect of his care and work. If you intend to move him from his present home, be prepared for his owner to want to vet you and the yard where he will be kept. You will need to draw up a formal loan agreement; many of the things covered in the section on sharing a horse will apply, but other areas to think about include the following.

- Is it to be purely a loan arrangement, or a period of loan with a view to buy? If the latter, agree a purchase price now, subject to a satisfactory veterinary report, so there is no possibility of misunderstandings arising.
- How long is the loan agreement for?
- Is the owner entitled to visit the horse at any time without warning?
- If you find you have to end the loan agreement, what period of notice should, in theory, be given? You may find that if you can't or don't want to keep the horse, the owner will take him back immediately, but in most circumstances it is fair to give an agreed notice period – and, of course, to look after the horse correctly during that time. Expect the owner to reserve the right to take the horse back if for any reason he or she feels there is a problem with the way he is being looked after.
- Is the horse already insured by the owner and, if so, does the policy cover him whilst he is on loan? If this is the case, make sure that you have a copy of the policy. If he is not insured, can you take out cover, especially for veterinary fees? This can be a problem with older horses, as many insurance companies are unwilling to cover horses beyond a certain age. In this case, you have to decide who is responsible for veterinary fees and up to what level.
- Hopefully nothing will go wrong and the horse will stay healthy, but you have to build in a worst case scenario. Make sure you have the owner's written permission that if a veterinary surgeon advises that the horse should be put down due to accident or illness, you are able to authorize this.
- Does the horse have any specific management needs, such as a particular type of feed or bedding or remedial shoeing?
- Make sure you know his veterinary history. For instance, has he ever suffered from sweet itch or laminitis?
- What activities are you allowed to do with him and are there any that are not allowed? For instance, would a previous injury make him unsuitable for jumping?

Buying

If you decide to buy a horse rather than loan or share, there are factors to take into account before you even start thinking about what sort of animal would suit you. There are three established ways of buying: from a private seller; from a bona fide dealer; and at auction. All have pros and cons. There may also be situations where you find you are looking at a horse sold through an agent.

Recently, there have also been cases of sales made via internet auctions, or payment being made on the basis of photographs and details posted on internet sites without the buyer seeing or trying the horse. It can't be emphasized enough that whilst internet sites can be an excellent way for vendors to advertise their horses, you should never consider buying a horse you have not seen or tried.

Trade or private?

There is a distinct difference in law between a dealer, who sells by way of trade or profession, and a private seller. Don't fall into the mistake of thinking that a private seller is automatically going to be a safer or better person to buy from than a dealer. For a start, in the UK, buying from a dealer will give you more protection under the Sale of Goods Act, which stipulates that goods must be of merchantable quality and fit for the purpose for which they are sold. Although it may seem callous, in law, horses are classed as goods in the same way as cars or washing machines.

Because a good dealer relies on his (or her) reputation, it is not in his interest to sell you an unsuitable horse. If you are pleased with your purchase, you will recommend him to other people and, if later on you decide to sell your horse or pony because he has been outgrown or you have become more ambitious, would probably think of asking the dealer if he has a bigger or more competitive replacement.

Also, a professional seller will be used to matching up horses and riders and should be able to steer you away from animals who are unsuitable – perhaps because your ability doesn't yet match your dreams – and in the direction of one who will provide you with fun and enjoyment. You may also be able to see more than one potential buy on the same yard, which can save on time and travelling.

Another consideration is that if you buy a horse from a dealer and find that you don't suit each other, most professionals will take the horse back within a certain time of purchase and find you another instead. Hopefully, that situation won't arise if you follow the advice in the next two chapters but, despite what the law says, horses are not machines and sometimes, despite the best efforts of all concerned, partnerships don't gel.

You won't, unless you are very lucky, get a bargain if you buy from a dealer. However, on the best yards you should find horses that are accurately described and offered for sale at a fair market price.

Most reputable dealers are proud of their reputation and trade under their own names or under a recognized business name. Unfortunately, there are also people who sell horses regularly but masquerade as private sellers; the owner who advertises horses for sale because of moving house or becoming pregnant for the sixth time that year is a bit of a giveaway, but others seem to believe that 'turning over the odd horse' still allows them to retain the status of a private seller. If in doubt, look elsewhere, because a seller who is economical with the truth in this aspect is likely to be equally economical in others.

A private seller is one who does not sell by way of trade or profession and, although you still have some protection under the Sale of Goods Act, it is not as widespread as when buying from a dealer. If the horse is for sale for good reason and for no fault of his own – perhaps a pony has been outgrown, an adult rider is giving up because of work or family commitments or a young adult is going into further education – then you can often find a nice animal at a reasonable price. Hopefully, the seller will also be anxious that the horse goes to a good home and will give you the animal's full history.

Unfortunately, private sellers are not always as good as professional ones at assessing and valuing horses. There is sometimes a tendency to see a horse advertised at £X and assume that because their horse is of similar type and age, it must also be worth the same amount even if it does not have the competition record or potential to match.

Sometimes, owners who have to sell a horse but do not have the facilities or the inclination to show him off to potential buyers will ask a professional to sell him on their behalf. This must be made clear when inquiries are made and adverts will often contain phrases such as 'Sold on behalf of a client.' In this case, the person carrying out the sale is acting as an agent for the vendor and, if you buy the horse, you may find you complete the transaction without meeting or speaking to the owner.

Buying at auction

On the continent, buying at auction is an accepted way of finding a horse and, in Germany and Holland in particular, there are auctions specializing in top-class competition horses. In the UK, auctions (or sales, as they are commonly called) are the standard way of buying and selling racehorses, but it is not an accepted way of trading in 'general purpose' horses and ponies.

Buying at auction is, in any case, an avenue that should only be considered by the very experienced. You do not get the same opportunity to assess and try a horse as when you are able to see it on its home ground and, in some cases, it is not possible to have a horse inspected by a vet before you make payment. Sales descriptions and warranties, when offered, are something of a minefield and you may find that if you buy a horse that is unsuitable or unsound, depending on the terms and conditions of the sale, you may not be able to return it.

The best advice when looking for a horse is always *caveat emptor* ('buyer beware') and this probably applies particularly to buying at auction.

your perfect partner

In this chapter you will learn:

- **how to work out priorities**
- **what are the characteristics of breeds and types**
- **how to assess conformation and movement.**

Before you can find the horse of your dreams, you need to differentiate between fantasy and reality, which isn't always easy. Most of us can imagine ourselves soaring over fences on a magnificent show jumper, or performing intricate movements on a top-class dressage horse – but the fact remains that even if we could afford a horse of this calibre, we wouldn't be able to ride it successfully. It would be the equivalent of giving someone who had just passed a driving test the keys to a Ferrari, only in this case, the Ferrari would have a mind of its own!

However, you can still aim to find a horse who will become a perfect partner, which in itself is a dream come true. Getting there takes a lot of thought, research and analysis, but if you divide your quest into key areas, you will soon have a blueprint to work to. The horse you end up with may not match it entirely, because every horse is an individual and there will be room for compromise – for instance, you may decide to look for a gelding but find a mare who seems to answer every requirement.

Some of the key points to consider in building up your blueprint are height, age, gender and breed or type. When you're looking for a first horse or pony, bear in mind that you are looking for an animal that will suit you now, not one that you will be able to make the most of in a couple of years time.

You might be lucky and find a lovely first horse who turns out to have hidden talents, but it's more likely that you'll find you've bought more than you can cope with at that stage. It may seem wrong to think about selling a horse before you have even bought one, but you and your horse will, in the long run, be much happier if you focus on the short term and accept that most riders need different horses at different stages of their riding careers.

It can be very difficult to part with a horse when you have become fond of him, which inevitably happens when you build a good relationship. However, finding him the right home to move on to if you get to the stage where you need a horse with different attributes is part of being a responsible owner – and it's far more responsible to do that than to buy a horse who turns out to be more than you can cope with and who you have to part with after a short time.

Although financial considerations will hopefully come second to the horse's welfare, a sound, well-mannered horse suitable for a novice rider will always find a buyer. On the other hand, a horse whose way of going has worsened because of poor riding and handling – which can happen in weeks – can diminish in value because he will need an often lengthy period of re-schooling by an experienced rider.

Height

In recent years there has been an increasing fashion for big (meaning tall) horses even when the rider would be just as happy with, and perhaps better suited by, a smaller animal. Height in itself is not an overriding criterium, as the horse's weight-carrying capacity also has to be considered.

A rider up to 1.78 m will be ideally suited to a 15–16-hh horse, which will stand between 1.52 m and 1.63 m at the withers and a 16.2 hh horse will be tall enough for just about anyone, provided it is capable of taking the weight. Weight capacity depends on the horse's build and conformation; a substantially built small horse will be capable of carrying a heavier rider than a taller, much more lightweight Thoroughbred.

Adults may also want to consider native ponies, which are very versatile and capable of more than many people realize. Some breeds can carry 89 kg or even more.

Smaller horses are not necessarily better than bigger ones, but don't fall into the trap of thinking that a big horse is going to be more talented or will make you look more impressive. A smaller animal may also offer more practical benefits, as it will usually be cheaper to keep. Some people believe that exceptionally big horses of 17 hh and over are more likely to develop soundness problems and to have shorter working lives, but much depends on the way they are looked after and their workloads.

If you are looking for a child's pony, ask your child's instructor for advice. Rather than buying a pony that is a bit too big in the hope that the young rider will 'grow into it', look for one that is the right size at present. If a pony is too big, your child may have difficulty riding him or even controlling him and this can soon lead to a child who doesn't want to ride any more. Although it might seem a difficult prospect, it is better to buy a pony that will give your child fun and pleasure for a year or two and, when he is outgrown, go on to do the same for another child whilst yours graduates to the next model up.

Age

Horses usually start their ridden careers at three or four years old and, if this is continued successfully, can be expected to have an all-round education and experience of life at around seven years old. Some breeds mature physically at a faster rate than others; for instance, Irish Draught horses and native ponies are often not fully mature until they are about seven.

Although there are exceptions, the generally accepted yardstick is that a novice owner and a novice horse do not make a good combination. Even if you are an experienced rider in terms of what you have achieved and are capable of doing when riding a school horse, being responsible for every aspect of your own horse's care, exercise and schooling is a very different scenario. This means that unless a young horse has been exceptionally well educated and has an excellent temperament, it's better to think in terms of one who is seven or older as your first one.

The prime age for an 'ordinary' horse is often said to be between seven and ten years old, but this also makes it the most sought-after age group and therefore the most expensive. Do not discount a horse in his teens or even older, provided he is sound, as he will still have a lot to offer. Ponies have even longer working lives and many thrive and compete into their twenties.

There are two possible disadvantages in buying a horse over 15 years. One is that you have to accept that you may be taking on the responsibility for providing his last home – and, by definition, either providing him with a retirement in which his care and welfare are not compromised or taking the final decision to have him put down. The other is that many insurance companies will only provide limited cover for older horses.

Gender

There is an old maxim that a rider can tell a gelding what to do, but has to ask a mare and discuss it with a stallion. It's amusing but, allowing for the fact that horses are as individual as people, generally true.

Stallions are not suitable for novice owners under any circumstances. They are not wild animals, but retain their natural instincts and behaviour and, as such, need to be in the care of people who understand them and can build mutual respect. Inevitably, your choice will be between a gelding (castrated male) and a mare.

Geldings are, in general, easier to deal with because they are not affected by breeding impulses, though a gelding that is castrated later in life may retain some stallion characteristics. Many mares exhibit changes in behaviour, ranging from slight to extreme, in the run-up to and during the periods when they would be receptive to mating. As they come into season roughly every three weeks during the breeding season, this may affect a large part of your spring and summer.

Mares, like stallions, are entire and whilst some riders are able to form excellent relationships with them, others find them too complicated. Unfortunately, if you buy a mare during the winter period when her breeding cycle is dormant, you will not know how she will behave when it starts and if you don't know her, will have to accept the vendor's assurance of her behaviour during that time.

Colour

Another old saying – the horse world is full of them – is that a good horse is never a bad colour. This can be interpreted two ways and the most sensible is that if you find a horse that suits you in every way, it doesn't matter whether he is bay, grey or any other hue!

However, everyone has individual likes and dislikes and if you really dislike a particular colour, it will prejudice you against a horse no matter how nice he is. One myth that must be discounted is that horses of a particular colour share particular characteristics – in particular, that chestnuts are 'hot' and fiery. Some people still believe that a chestnut mare is the worst combination in the world, but that is totally unfair and there are as many good chestnut mares as of any other colour.

Fashion has a large part to play. For instance, at one time skewbald (see Plate 3) and piebald horses – often known collectively as coloured horses – were looked down on and thought to be suitable only for travellers. Now, they have become incredibly popular and a good coloured horse will often fetch a higher price than an equally good one of a solid colour.

Breeds and types

As you start your search for a horse you will find that some are of known breeding and some are described of being as a particular type, such as a cob type. A cob (see Plate 4) is a deep-bodied, up-to-weight horse whose legs are often relatively short in proportion to its body compared with, say, a Thoroughbred. Cobs usually have a percentage of draught or heavy horse blood and often have good temperaments and willing attitudes that make them ideal for novice owners as well as being fun for more experienced ones. A cob is a type rather than a breed, but there is one exception – the largest of the Welsh breeds, the Section D, is also referred to as a Welsh Cob.

When a horse has characteristics that hint strongly of a particular ancestry, such as Connemara or Irish Draught, but no breeding documents, it may be referred to as a Connemara or Irish Draught type. Some sellers also think this is a good way to interest buyers in popular breeds, so don't assume its accuracy!

Another common description is the 'ride and drive' type. This denotes a horse who is either broken to drive as well as to ride, or one who the vendor thinks would be suitable for both purposes.

There is nothing wrong with buying a horse of unknown breeding, and a suitable horse of unknown parentage is a better buy than an unsuitable one with blue blood ancestry, but different breeds have definite physical characteristics. Whilst temperament is to a large extent down to the way a horse has been trained and handled – and he will be the sum of his experiences, whether good or bad – some breeds seem to have similar traits.

There are many types of breed throughout the world and whole books devoted to them. Below are some of the ones you are likely to see in the UK, Europe and the USA. Many horses will be proven crosses, usually with the Thoroughbred – for instance, the Thoroughbred cross Irish Draught and Thoroughbred cross Connemara are very popular.

Arabian

The Arabian horse, sometimes called the Arab (see Plate 5), is famous for its stamina and usually stands between 14.2 hh and 15.2 hh. Although they are lightweight in build, they can take a heavier rider than their appearance suggests. They have naturally high head and tail carriage and their heads have a dished or concave profile.

These horses thrive on work and are usually intelligent. They can be sensitive and quick to react and might not be the first choice for a novice owner. Over the years they have been crossed with many other breeds, from Welsh ponies and Welsh Cobs to Thoroughbreds. Of all the breeds, the Arabian and the Thoroughbred have had the greatest influence in shaping today's horses and ponies.

Anglo-Arab and partbred Arab

The Anglo-Arab is half or three-quarters Thoroughbred, the remainder being Arabian. The partbred Arab has a minimum of 25 per cent Arabian blood and the rest can be any breed except Thoroughbred on its own – though some Thoroughbred may be included.

In theory, such breeding produces the best of all the components. In practice, one may tend to dominate, with varying results.

Cleveland Bay

The Cleveland Bay, which is always bay in colour, was bred to be a driving horse and still takes that role. It is now officially a rare breed, as it is usually crossed with the Thoroughbred to produce a riding horse and the number of purebreds has decreased. Purebred Cleveland Bays are usually 16.2 hh or above.

Irish Draught

The Irish Draught was bred originally, as its name suggests, as a workhorse in Ireland. Today, it is most frequently crossed with the Thoroughbred to produce a riding horse known for its jumping ability and good temperament. Most of these crosses will be 16.1 hh and upwards.

Quarter Horse

The Quarter Horse is an American breed that has become popular in the UK, especially with devotees of Western riding – a form of riding which is becoming hugely popular. It gets its name from the fact that it was raced over distances of a quarter of a mile.

Quarter Horses are usually between 14.2 hh and 15.2 hh, with laid-back temperaments that make them easy to handle.

Thoroughbred

The Thoroughbred (see Plate 6) and the Arabian are the most influential breeds in the world. The Thoroughbred is bred for speed and was bred first and foremost to race, but it also excels in top-level eventing because of its speed and stamina.

Thoroughbreds bred for flat racing are smaller and lighter than those bred for racing over jumps, but all have elegance and clean lines.

Thoroughbreds are usually quick to react and may not be suitable for novice owners. When they are crossed with another breed, their offspring often does have the best of both worlds. Such importance is given to the Thoroughbred portion that a horse who is half Thoroughbred is known as a halfbred and one who is three-quarter Thoroughbred as a three-quarter bred (see Plate 7).

Warmbloods

Warmblood is an overall term used to describe horses produced through breeding programmes, originally on the continent, including Germany, Holland, Denmark and Sweden. Germany and Holland in particular have selective breeding programmes that can be traced back for many years, allowing them to predict the characteristics and abilities of certain lines. There are now many good British warmbloods (warmbloods bred in Britain).

The modern warmblood is usually three-quarter Thoroughbred or even more. They dominate top-class competition in dressage, show jumping and eventing.

Pony breeds

Connemara

The Connemara (see Plate 8) is probably the most popular native pony and can do any job. It is also suitable for adults as well as children and a good, registered Connemara will fetch a high price. Heights range from about 13 hh to 14.2 hh – the breed society limit is 148 cm, which is just over 14.2 hh. Some grow over the permitted height, but are just as much in demand.

Dales

The Dales pony is the real weight carrier of the pony world and in the First World War carried loads of up to 130 kg. It is strong, hardy and, although heavily built, is light on its feet with a powerful trot. Dales ponies are usually 13.2 hh upwards, with most being around 14.2 hh.

Dartmoor

Dartmoor ponies have smooth, free movement and make excellent riding ponies for children. Their maximum height of 12.2 hh means that most adults will consider them too small even though they can carry more weight than their size suggests.

Exmoor

The Exmoor is the oldest of the British pony breeds (see Plate 9). They are incredibly strong and hardy and are able to carry up to 70 kg. Again, their maximum height of 12.3 hh would make all but the smallest adults discount them.

Fell

The Fell pony has many characteristics in common with the Dales, such as hardiness and lots of feather (silky hair at the heel). It is usually slightly smaller – the breed society states a maximum height of 14 hh – but is also up to weight and suitable for adults.

Haflinger

The Haflinger is an Austrian breed that has become very popular in the UK and Europe. It is always a shade of chestnut, ranging from light to deep, with a light-coloured mane and tail. Its height ranges from 13.2 hh to 15 hh and, like the larger British breeds, a good Haflinger will suit most members of the family.

Highland

Highland ponies (see Plate 10) were bred to carry deer carcases weighing around 100 kg and, whilst a Highland wouldn't be the speediest of mounts, it is a popular one for adults who want a weight carrying animal. Their maximum height is 14.2 hh.

New Forest

This is another breed that is excellent both for children and small adults, as it ranges in height from 12.2 hh to 14.2 hh and is deep bodied enough that a long-legged adult would not look out of proportion on a pony over 14hh. In their native habitat, they are ridden by the agisters (wardens) as they monitor and look after the New Forest.

Shetland

Everyone knows the Shetland, the smallest breed of pony and the only one measured in inches rather than hands high. Its height, or lack of it – around 96–102 cm – means it is only suitable for children to ride, but it is very strong and makes a great driving pony. Unfortunately it is often bought as a pet or a lawnmower on four legs and unless it is handled correctly, as with any other horse or pony, it will inevitably be badly behaved. Most problems with Shetlands are not the fault of the ponies, but the people who own them!

Welsh

There are four Welsh breeds, divided into Sections A, B, C and D. Section A and B ponies are the smallest and are deservedly popular children's mounts; the Section C goes up to 13.2 hh and is suitable for older children and adults and the Section D is usually around 15 hh.

All the Welsh breeds have a 'look at me' quality. The Section D, also called the Welsh Cob (see Plate 11), has powerful movement and has a reputation for being forward going and even fiery; as with all reputations, some live up to it and some don't.

Conformation

Whatever type or breed of horse you decide to buy, he should have good enough conformation for the job you intend him to do – in other words, he should be put together in a way that allows him to work efficiently and stay sound. Obviously you would not expect a quiet, perhaps elderly pony who is intended as a first mount for a small child and will not be expected to gallop or jump to meet the same criteria as a potential top-class competition horse, and in the case of the former, handsome certainly is as handsome does. But for most riders, it's important to find a horse without too many 'weak links.'

Conformation is not just a pretty picture, it is a blueprint for biomechanical efficiency. At the same time, there is no such thing as the perfect horse, so it is a case of assessing an animal's good and not so good points and deciding whether one outweighs the other. If you aren't sure of your own ability to judge conformation – and it really only develops by looking at many different horses and discussing them with those who are more knowledgeable – you can get someone experienced to help you. But as you are the one who is going to make the horse part of your life, it's important to hone your own judgement.

Use the following guidelines to take a good look at horses you know and, ideally, have ridden – and remember that one of the most important parts of a horse's conformation is the part you can't see. It's the bit between the ears: in other words, his temperament!

First impressions

Start by forming an overall impression of the horse's symmetry. Does he seem to be built in proportion, or does it look as if the front end doesn't match the back? Do his limbs seem suitable for his body, or are you looking at a chunky, deep-bodied frame on spindly little legs? Are the proportions of his head and neck in keeping with those of his body, or does he have a long back and a short neck? Once you have built an overall picture, you can focus on its individual components.

Feet and limbs

Don't start with the head looking over the door, because although it will count towards your overall impression, it won't help keep the horse sound. Instead, begin from the ground up: there is an old saying, 'No foot, no horse,' which translates as a warning that if the horse has badly formed feet, he won't stand up to work no matter how beautiful the rest of him is. A good farrier can do a lot to help, but won't be able to work miracles.

Does the horse have two matching pairs of feet? Hind feet are usually slightly smaller than front feet, but each pair should be the same size and shape. Are they in proportion to the rest of him? Feet that are a bit too big are better than feet that are much too small; you don't want to see a big, substantial horse with feet that look as if they belong on a small pony.

When you pick each foot up and look underneath, the soles should be concave rather than flat, as flat soles are more prone to bruising. Heels that are too low are also a weakness, though you may need a farrier or vet to determine whether this is poor conformation or due to bad or too infrequent shoeing.

When you look at the angle at which the hoof joins the limb, you want to see a continuous slope from the pastern down to the ground rather than a 'broken' angle. The cannon bones should be short rather than long and should have a straight profile from the bottom of the knee to the top of the fetlock.

If the profile is slightly concave, he is said to be back at the knee; if it is convex, he is over at the knee. Both are defects because they can put extra strain on the tendons that run down the back of the leg, but usually only cause problems if they are marked.

Riders who need to make sure their horse will be able to carry a certain amount of weight easily will also need to assess how much bone the horse possesses. This peculiar phrase denotes the circumference of the cannon bone at its widest part, just below the knee. Bone and conformation should go together – a cob or substantial halfbred type should have at least 22–23 cm (8.5″–9″) of bone, whilst a small, lightly built Thoroughbred may only have 18 cm (7″). As with so many things in the horse world, these measurements are traditionally made in imperial rather than metric, so it helps to be numerically bilingual!

However, working out weight capacity isn't just a case of number crunching. If the foot and limb conformation is good, it may outweigh the fact that the bone is theoretically insufficient – usually described as the horse being 'light of bone.' It is important to look at the overall picture, so if in doubt, get expert advice.

The forearm, the part of the limb above the knee, should be long in relation to the cannon bone. The pasterns should be reasonably long, but not exaggeratedly so – if the pastern is too long, it puts too much wear and tear on the fetlock joints. All joints are important structures, because they act as levers and shock absorbers, so look to see if knees, fetlocks and hocks match their partners.

At the same time, pasterns that are too upright – a type of conformation that often goes with upright shoulders – will not give as comfortable a ride, because they will not be able to absorb concussion as efficiently.

Hindlegs are as important as forelegs, because the back end of the horse is his engine. One of the worse faults is a hindleg that is too straight as it runs down to the hock, because flexion is impaired and too much strain put on the fetlock joint.

When you look at a horse from the front, he should not look as if his cannon bones are offset to the side of the knee. Nor should his feet turn out at 'ten to two' or turn in, so that he is pigeon toed. From behind, his hocks should not turn markedly inwards like those of a cow – a fault called, not surprisingly, 'cow hocks'.

Whilst it's important to notice any defects, it's also important to keep reminding yourself that if you go on looking for the perfectly made horse you'll never find him. Imperfections are not important in isolation: it's their degree and their relationship to other structures that count.

Limb blemishes and defects

You'll often hear people talking about the importance of 'clean limbs', which has nothing to do with whether or not they are covered in mud. This phrase means that he does not have any lumps or bumps. Unless you are buying a horse to compete in top-class showing classes, blemishes may only be important if they affect the horse's soundness or are the sign of an exaggerated weakness in limb conformation that could lead to further problems.

It is also inevitable that horses, like people, show signs of wear and tear as they get older, particularly if they have led active lives. Your vet will explain any blemishes and their significance during a pre-purchase veterinary examination, but the commonest ones are explained below.

- Splints – bony growths on the inside of the cannon bone, usually on the forelegs, caused by wear and tear or occasionally by a blow. They form as the skeleton's defence to concussion and, although the horse may be slightly lame when they are forming, they rarely cause problems once they have settled unless they are in a position that interferes with the action of the knee joint.

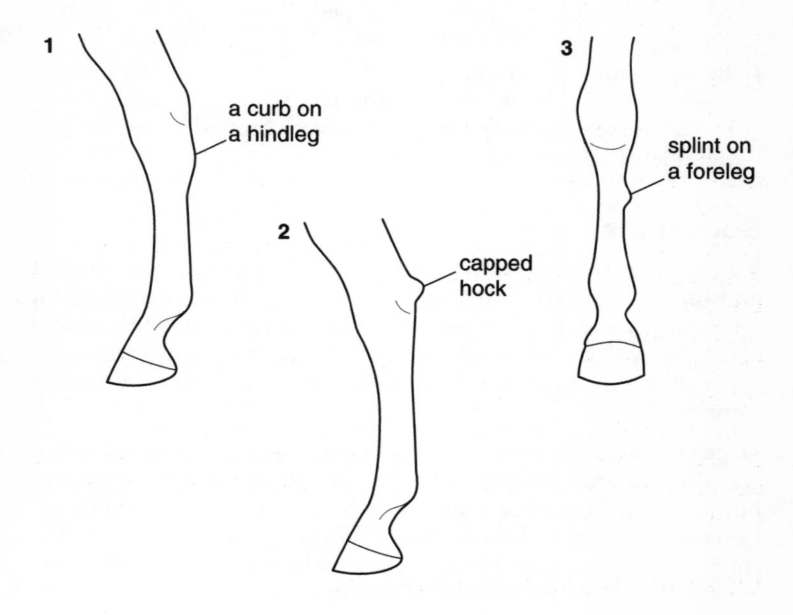

figure 11 splint, curb and capped hock

- Windgalls – fluid-filled swellings below the fetlock joint that are usually painless and cause no problems.
- Bog spavins – where the natural depression in front of the hock appears to be filled in. Usually, these do not cause a problem. They are not to be confused with bone spavin, which are changes in the bones of the hock joint which only show up on X-ray and usually cause lameness.
- Curbs – swellings on the back of the hindlegs just below the hock, caused by ligament strain. Unsightly but rarely a problem unless found in conjunction with poor hindlimb conformation.
- Thoroughpins – swellings just above the hock joint; blemishes but rarely result in lameness.
- Capped hocks – blemishes that are usually caused by the horse scraping his hocks on bare stable floor when he gets up, perhaps because he has insufficient bedding. These soft, fluid-filled swellings on the point of the hock are unsightly but have no affect on soundness. Capped elbows are less common but, again, are really only a cosmetic consideration.
- Bowed tendons – swellings on the back of the leg caused by a tendon injury. An injured tendon is always a weakened one, so the horse's capacity for work may be limited. Your vet will be able to advise you.

Body lines

Whilst most people put the greatest importance on a horse's feet and limbs, his body shape will also affect how easy or difficult he finds it to work. For instance, if he has a reasonably broad chest, he will be less likely to knock one front leg against the other than the horse who has 'both front legs coming out of the same hole.'

Depth through the girth – not to be confused with the expanded waistline of a fat horse – is another attribute for a performance horse, as it means there is plenty of room for his heart and lungs. Look at the length of the horse's back, too; one that is markedly short will make saddle fitting difficult whilst a back that is overly long will tend to be weaker. Mares are naturally slightly longer in the back, as they need extra room to carry a foal.

The angle of the shoulder should have a marked slope if the horse is to be a comfortable ride. If it is too upright, he will tend to have an 'up and down' action rather than movement which flows forward.

Hindquarters need to be able to generate enough power. The horse who looks as if his quarters belong to a much smaller or more lightly built animal will probably be weak in this area, though it sometimes takes an experienced eye to distinguish between a good skeletal frame that lacks the correct build-up of muscle and a back end that will always be weak.

When the horse stands on flat, level ground, he should not look as if he runs downhill from his hindquarters to his withers, even if he is of a breed or type with naturally low withers, such as a cob. The downhill horse will naturally carry more weight on his forehand, so it will be more difficult to train him to carry himself in a balanced way. Although you are unlikely to be looking for a young horse who has not finished growing, remember when you are comparing different horses in your study of conformation that horses do not grow evenly: they can go up at one end and down at the other in an alarmingly see-saw growth pattern until they mature.

Head starts

The shape of a horse's head is usually the last thing to worry about, unless you feel you really couldn't bear to see him looking over your stable door every morning! Some heads are pretty, some plain and workmanlike. As you can transform a plain head into a handsome one by using a bridle with a broad noseband, it shouldn't be a priority.

A lot of importance is often placed on a horse having a kind or generous eye, which means a large eye. Whilst it is attractive, it doesn't follow that a horse with a small eye will have a less generous temperament. Human faces don't always meet the accepted standards of beauty, and neither do equine ones.

Similarly, whilst a horse who is frightened or aggressive may show the whites of his eyes whilst exhibiting this behaviour, some horses have a natural white ring around one or both eyes. In terms of temperament, it means nothing; nor do light-coloured eyes, often called blue or wall eyes.

What is worth noticing is the way the head is set on to the neck and the angle at which the neck comes out of the withers. If the angle of the throat is a sharp V rather than a curved U shape, the horse may find it more difficult to flex and carry his head correctly. However, a lot of cobs have this fault and, unless it's really noticeable, the horse should still be able to work nicely.

A fault which should be avoided, if possible, is a ewe neck. This is when the neck appears to have been set on upside down: such horses find it hard to work correctly and are often stiff through the back.

The mouth and ageing

Looking in a horse's mouth can tell you two things. One is whether he has normal jaw conformation and the other is his age – hence the advice not to look a gift horse in the mouth.

Holding the horse's lips gently away from his teeth will enable you to see if he has normal jaw conformation or is markedly overshot or undershot. An overshot jaw is when the upper jaw is too long; it is commonly called a parrot mouth, because of the resemblance to a parrot's beak. An undershot jaw is the opposite condition, when the upper jaw is too short. Minor misalignments should not cause problems, though regular dental care is particularly important. If the condition is marked, the horse may find it more difficult to graze and chew.

An expert can get a good idea of a horse's age by looking at the number and appearance of temporary and permanent teeth, though it is now recognized that this is not totally accurate and can be affected by factors such as heredity and environment. It is possible to get a reasonably accurate idea of the horse's age up to eight or nine years, but after that it becomes more difficult.

Movement

Straight movement – which basically means that the horse's limbs and feet follow an ideal flight path on each stride and that he does not throw his legs inwards or outwards – is usually dependent on good conformation. Deviations from straight movement are often caused by limbs and/or feet that are offset in some way or by other conformation faults.

Straight movement puts less strain on the limbs and so helps a horse to stay sound. You may also hear a horse described as an extravagant mover, which doesn't mean quite the same thing; an extravagant mover is one with a long stride and also usually implies that, with each stride, the horse spends relatively more time off the ground. Extravagant movement may be a desirable quality for a dressage horse and some types of show animal, but isn't necessary for the all-round horse. Unless the rider is very balanced, a horse with very extravagant movement can be more difficult to ride.

There are four common types of movement that are technically faulty: brushing, dishing, forging and plaiting.

Brushing means that the horse knocks one leg against its partner, either in front or behind. It can be caused by faulty conformation, when the legs are too close together – as with a narrow-chested horse – or weakness. Protective boots may be sufficient to manage mild or temporary cases, but a horse who brushes badly is likely to injure himself.

Dishing is when the horse turns out one or more front feet to the side, which puts more strain on the limb or limbs. Its significance depends on the degree of severity; a horse that turns a toe slightly should not have any problems and should certainly not be discounted if he suits you in all other respects. A horse that dishes badly, usually because the limb is not set on straight from the knee or fetlock, may be more susceptible to wear and tear, and if you really like him in every other way, you need to ask your vet's advice.

Plaiting, where the horse crosses his front legs as he moves, is less common. Although you would think that a horse who moves in this way would be in danger of tripping himself up, most actually manage not to do this. As with any deviation in movement, it puts strain on the limbs.

Forging describes the action of a horse who strikes the back of the front foot with the toe of the hind foot; you can often hear the sound of one shoe striking another and, in some cases, this may result in an injury to the heel called an overreach. Unless the horse is particularly short in the back, forging is often caused by lack of balance and will often improve as the horse strengthens up and learns to carry himself in balance.

Stereotypical behaviour

Some horses find being confined to a stable stressful and follow habitual behaviour patterns that are now thought to be a sort of nervous defence mechanism, as discussed briefly in Chapter 03. At one time this was commonly known as stable vices, but it is now accepted that this term is unfair and many people use the term stereotypical behaviour instead. The argument is that horses are by nature animals who live outdoors and that the horse behaves in this way because it is kept in a way contrary to its natural instincts.

It is wrong to assume that all horses dislike being stabled; some are relaxed about it and will demonstrate quite clearly that when it is cold, windy and raining they would far rather come in than stay in the field. However, it does demonstrate the need for horses to spend a large part of their time outdoors on well-managed land with adequate shelter.

To recap, the three main recognized forms of stereotypical behaviour are weaving, crib biting and wind sucking and all should be declared when a horse is sold, as they can affect the horse's health if severe. Some horses show the behaviour all the time and others only under certain circumstances: for instance, when they are about to be fed and when they are moved to a new home.

A horse who is a weaver moves his head from side to side over the stable door. In really bad cases, he may shift his weight from one front leg to another, which over time may put unwanted strain on his joints.

A crib biter sets his teeth on a covenient place – usually the top of a door, a ledge in the stable or a fence rail – and bites down on it. This causes wear on the teeth and to whatever he is crib biting on. Wind sucking, where the horse gulps in and swallows air, is often practised in conjunction with crib biting, but some horses will wind suck without needing anything to set their teeth against. It is potentially the most harmful of the behaviour patterns, as it may predispose a horse to colic.

Box walking, where the horse walks round and round his stable, must be considered a form of stereotypical behaviour, but for some reason is often not counted in with the 'big three.' Again, it puts unwanted wear and tear on the horse and, because he is constantly on the move, in some cases it may be difficult to keep weight on him.

Attitudes to stereotypical behaviour vary. Some people say that if the horse is excellent in all other respects – and particularly if he is talented – then it is of no consequence. Certainly few riders would turn down a potential Badminton winner or Grand Prix dressage horse because he weaved. But with the 'ordinary' horse, which is what most of us own, it is usually a disadvantage. Some yard owners will be unwilling to accommodate horses who exhibit stereotypical behaviour, usually because they or other clients are worried that other horses on the yard will start copying it. In fact, researchers now believe that horses who copy behaviour seen in another would already be predisposed to it.

Unless a horse is mega talented, stereotypical behaviour will lower his value. It may also make him more difficult to sell if the need should arise, so you have to decide whether or not it would be acceptable.

You may be able to keep him in a way that would minimize or in some cases, eliminate the behaviour. For instance, few horses weave when kept out of doors, so if you are able to give him a 24/7 outdoor lifestyle, you may feel you can afford to ignore it. Horses who crib bite and wind suck indoors will often do so outdoors; it is now believed the behaviour may be associated in some cases with gastric ulcers and the use of antacids has been effective with some horses.

At one time the accepted methods of dealing with stereotypical behaviour centred on preventing them and some yards still follow this approach. A V-shaped grid on the stable door may dissuade a mild weaver simply by preventing him from moving his head from side to side – but there is also the risk that he will step back from the door and perform the same behaviour, whilst becoming more stressed.

One practice generally held to be unacceptable is the use of anti-cribbing collars. These are leather straps with U-shaped metal sections that are buckled round the horse's neck, so that when he arches his neck to crib bite and/or wind suck, the metal digs into his throat and makes it too uncomfortable.

In general, it is now believed that preventing stereotypical behaviour causes the horse more stress than allowing it. So whilst painting foul-tasting substances on the stable door may dissuade the horse from crib biting, he may become so stressed that he develops other problems.

12 finding, trying and buying

In this chapter you will learn:

- **how to find horses for sale**
- **how to identify a potential purchase**
- **assessing and trying horses**
- **completing a purchase.**

Horses for sale are advertised in specialist and local publications and on specialist internet sites; some sites run as independent concerns whilst others are linked to equestrian magazines. It is also worth looking on notice boards in local saddlery shops and feed merchants – and, of course, telling everyone you know in the horse world that you are looking for a horse. Instructors and retailers often know of clients and customers who have horses for sale and may also be able to point you in the direction of good dealers.

Rather than go and see every horse who sounds like a remote possibility, narrow your search by asking the right questions. This will save you time, money and hassle – looking at horses may seem exciting, but by the time you have spent the umpteenth weekend looking at the umpteenth unsuitable one your enthusiasm may have waned. It is also a waste of the seller's time, so don't be surprised if, when you start making inquiries, you find yourself answering as many questions as you ask. In fact, you can usually take that as a good sign.

When you read the adverts, your first impression may be to wonder why so many paragons of virtue are on the market. The reality is that whilst everyone selling a horse will want to emphasize its good points, a lot of horses described as 'stunning' or 'talented' don't live up to the sales pitch. This is where a picture can tell a thousand words; many advertisers include pictures of their horses and, as most people have access to e-mail, it is rarely worth going to look at a horse on the strength of a written advert alone. There is much a photo will not reveal, but you should be able to spot a horse in very poor (see Plate 13) or very good (see Plate 14) condition.

On the other hand, when you see horses for sale who appear to be as stunning and talented as the advert claims, remember the need to be realistic. Often this is quite easy, as the price of a top-class horse will be much higher than most novice owners want or perhaps can afford to pay.

Looking at adverts should give you an idea of market prices. Good dealers are another guide; they will usually price their horses at the top end of the range, but unless you give the impression that you have just won the Lottery and are anxious to buy your first horse, should not ask more than they can realistically expect to get!

You will also, of course, be restricted by your budget. Whilst there is often a certain amount of leeway between the price someone asks for a horse and what would be accepted, you again need to be realistic. Someone who advertises a horse for £4,250 may be prepared to take £4,000, but would be very unlikely to accept £2,500. So whilst it's worth inquiring about horses who are advertised at slightly more than you can afford, looking at ones who are vastly more expensive is a waste of everyone's time.

Another point to consider is how far you are prepared to travel, as time and travelling costs can be expensive. Against that, if lengthy discussions with the vendor – perhaps backed up by a video of the horse being ridden that you can watch and show to someone more knowledgeable, such as your instructor – persuade you that this horse could be the one, you may feel the effort and expense is worthwhile.

Understanding the lingo

There are many commonly used phrases and initials used in adverts. Some of the ones you are likely to come across are listed below.

- BD points/BD registered – the horse has won points through competing successfully in dressage at affiliated-level competitions run under the auspices of British Dressage, or is simply registered with British Dressage. However, the fact that a horse is registered doesn't in itself have significance, as it could be registered and have performed unsuccessfully or even badly when competed.
- BSJA winnings/registered – the horse is registered with the British Show Jumping Association/has winnings in affiliated show jumping competitions.
- CB – Cleveland Bay.
- Good doer – a horse who thrives on the minimum amount of feed and is therefore supposedly cheaper to keep.
- Halfbred – one of the horse's parents was a full Thoroughbred.
- HT – may be horse trials, or eventing, where the horse is required to compete in the phases of dressage, show jumping and cross-country within one competition; or hunter trials, where horse and rider tackle a course of cross-country fences.
- ID – Irish Draught.
- Lightweight, middleweight, heavyweight – denote the horse's type and also the type of rider he is able to carry. Often abbreviated to lwt, mwt and hwt.

- Lives in or out – a horse or pony who is happy to be out in the field or to be stabled.
- MM or M&M – mountain and moorland.
- Partbred – a horse or pony whose parentage contains a known percentage of a registered breed: for example, partbred Welsh.
- Ride and drive – a horse or pony who is broken to harness as well as a riding animal. The phrase 'potential ride and drive' means the seller considers it suitable for both jobs, not that it is already trained for both.
- SJ – show jumper.
- TB – Thoroughbred.
- WH – working hunter; usually implies the horse is competing in or would be suitable for working hunter competitions, showing classes where the horse is judged on its ability to jump a course of natural-looking fences as well as on its conformation and movement.
- WHP – working hunter pony; similar to above but with relation to ponies.
- XC – cross-country.

Questions to ask

When reading advertisements, you need to study not only what is said, but also what isn't. This gives you a list of questions to ask so that you can fill in the gaps between the sales description and your specific needs: for instance, if a horse is described as good to box, clip and catch, does this mean that he is not good to shoe, or simply that the vendor has forgotten to include the information?

It's a good idea to keep a list of questions handy as you inquire about horses. Ones which may be particularly useful include:

- Does the horse belong to the advertiser or is it being sold on the owner's behalf?
- If you are speaking to the owner, how long has he or she owned the horse? Is the vendor a private seller or a dealer?
- Has the seller measured the horse, or is the height declared a 'guesstimate'? Unfortunately, some people are wildly inaccurate and nothing is more annoying than travelling to see a 15.2 hh who turns out to be nearer 14.2 hh.
- Does the seller have proof of the horse's age? Although it's possible to tell the age of a horse up to about eight years old

by studying its teeth, dentition is not as reliable a guide after that. Breed registration documents incorporated with a passport are reliable as long as you are sure the documents belong to this particular horse – something a vet will check during a pre-purchase veterinary examination.

- Does the horse have a passport? It is now an offence in the UK to sell a horse without a passport that has been issued by a recognized authority. All equines, whether of known or unknown breeding, must have passports.
- Is the horse suitable for your purpose? Tell the vendor exactly what you want to do with the horse and don't exaggerate your ability or experience.
- Is the horse good to hack out alone and in company, in all sorts of traffic? If you are told that he 'prefers company' or 'doesn't always like lorries' be aware that this can translate to a horse who refuses to leave the yard on his own or is likely to try and whip round and gallop off at the sight of anything bigger than a van! If you want to hack out, and most riders do, reliability in traffic is something that should never be compromised on.
- How has the horse been kept? If he is used to living out day and night and you will need to stable him part of the time, you need to be sure that he will adapt.
- Does he weave, crib bite, wind suck or box walk, or has he ever shown signs of any of these behaviours? If you are told that, for instance, he may weave in a new home for the first few weeks but should then settle down, you have to decide whether or not to take a gamble that the prediction proves accurate. The seller has fulfilled the necessary obligations by telling you that the horse may or has been known to weave.
- Does he need any special care or management and does he have any allergies? It is good management to keep all horses on a regime that keeps their environment as free from dust as possible, but if a horse has known problems, can you manage him accordingly?
- Is the horse freeze marked (see Plate 12) or identichipped? If so, and it turns out that this is the horse you want to buy, you will be able to check that the horse is not registered as stolen. In the case of a freeze marked horse, check that he is not marked with the letter L in a circle, as this will mean that he has been the subject of an insurance claim for loss of use. If he has, be wary but do not necessarily discount him; for instance, a horse who is no longer suitable for competition may be perfectly suited for lighter work and you will need to discuss things with the vet who examines him for you.

- What sort of bit and bridle is he ridden in? Although this may sound odd, it will often give you a clue as to what the horse is like to ride. For instance, if he is always ridden in a simple snaffle bridle you will hopefully find that he is not strong or difficult. Don't discount a horse if the seller says he needs a different noseband or a bit such as a pelham for show jumping or cross-country, but do be aware that this means he is more onward bound in these situations.
- Is the horse good to load and to transport? If you intend to buy your own transport and will be opting for a trailer and towing vehicle rather than a horsebox, is the horse used to travelling in a trailer and is he happy doing so? Most horses travel well in trailers when given a good journey by a considerate driver, but a minority does not adapt – either because the horse has had bad experiences or because he prefers to travel facing sideways or backwards; this is not possible in most trailers, as the majority requires the occupants to face forwards.
- Has the horse had any veterinary problems whilst in its present ownership? If you decide you want to buy him, would the owner let you request a veterinary history from the vet who has provided routine care such as vaccinations? If the answer is no, you have to ask yourself if there is something the seller does not want you to find out!
- If you decided to buy the horse, would he be open to a pre-purchase veterinary examination? Don't consider looking at a horse if the owner would not permit this, as there has to be a reason.
- Last but not least, why is the horse for sale? It's often best to keep this question until last, as by then you should have formed an impression as to whether the seller is being straightforward or less forthcoming. If phrased correctly, it should not cause offence and there will hopefully be a genuine reason.

A seller who genuinely wants to find the right home for a horse will ask you just as many questions and honesty really is the best policy. When you find a horse you want to look at, make an appointment to go and see him and, unless you already know how to find the yard where he is kept, get detailed directions – including the name of the premises, the name of the road, the name and telephone numbers of the owner and a mobile number where necessary so that you can call and ask for directions if you get lost. Give your own details in return so that if the horse is sold before you are due to see him, the owner can let you know and save you a wasted journey.

Trying times

Most novice owners – and many more experienced ones – like to take someone with them when they go to look at a horse. A novice should certainly get the opinion of someone knowledgeable, such as an instructor; even if you are confident of your judgement, it's useful to have someone similarly experienced with you. If you are asking a professional such as your instructor to go with you, expect to pay for their time; you may want to go alone for a first visit and, if all seems favourable, return as soon as possible with your consultant.

It's polite to ask the seller if there is any objection to you bringing someone; there certainly shouldn't be and you may be wary of someone who knows you are inexperienced and tries to persuade you that you don't need any help. If you intend to video the horse so that you can show your instructor or advisor, it's good manners to ask beforehand if this would be all right. One thing you shouldn't do is arrive with a committee in tow – and if you are looking for a horse for yourself and have children, try to arrange things so that you don't have to take them with you. You can't concentrate on assessing a horse and look after children at the same time and you can't expect the seller to look after them for you.

When you arrange to see a horse, try and get there at the agreed time. It is sometimes suggested that you deliberately arrive an hour earlier to catch out an owner who might be doing something untoward, but this is not recommended. For a start, you may inconvenience the seller, which won't exactly get your meeting off on the right foot – and if he or she intended to work the horse before you got there to try and guarantee a quieter ride, it will have been done first thing and the horse will be back in his stable. However, it is a good idea to ask if the horse has been ridden that day, as someone might have been to see him before you.

The assessment guidelines suggested below assume that you are making a first visit alone and, if you like the horse, a second with an advisor. If you prefer to take someone with you right from the start, you will be able to carry out all the trial stages on one day. No matter how much you like the horse, it's usually still best to try the horse a second time to confirm your first impressions, or at least to allow time for you to discuss what you have seen with your advisor and be sure that you are making the right decision.

There is, of course, always the risk that another buyer will see the horse before you can return and agree to buy him. This can be very disappointing, but there is no such thing as only one horse for one rider and you will find another that you like as much or even more. It is also better to lose a potential buy than to make a bad one because you did not try him thoroughly enough.

The guidelines below are also targeted to a horse who will be ridden by a reasonably competent adult or young person. Whilst many of the considerations apply to children's ponies, there may be different priorities – especially when buying a first pony, as the main priorities here are that the pony is quiet, reliable and easy to handle.

Initial inspection

Most people will have the horse ready in his stable for you to see, hopefully well groomed so you don't have to try and visualize what he looks like under a layer of dried mud. Unfortunately, some people apply hoof oil because they think it adds to the overall smartness, which can make things unnecessarily messy when you pick up the horse's feet – but could it also be an attempt to disguise a problem, such as crumbling hooves?

Be careful, too, if the horse is tied up in his stable or standing tied up outside it. If this is the case, ask to see him untied in his box. It may simply be that you have arrived as the owner finished grooming him, but it could be an attempt to hide the fact that he displays stereotypical behaviour, lays his ears back and pulls faces or even nips. Some horses pull faces because they regard their stable as their territory, or because they have been annoyed by incorrect handling; if that is all it is, you may feel able to ignore it, but be aware of the horse's attitude and behaviour when he is brought out of his stable.

Whilst it's nice to see a horse looking over the door with a calm, alert expression and a willingness to let you greet him, don't worry if on first inspection he looks half asleep. He's on his home territory and may be a laid-back character who doesn't expend his energy when he doesn't have to. The only time to worry is if he doesn't want to expend it at all!

With the owner's permission, go into the stable and approach the horse quietly. Spend a little time talking to him and run your hand down his neck. Ideally, he should seem calm but interested.

If he is wearing a rug, the seller should take it off so you can inspect him close up. This gives you a chance to run your hands over him, gauging his reactions and noticing any lumps, bumps or areas which seem uncomfortable. Does he tense or dip away from pressure in the saddle or girth area? Is his musculature even on both sides or is one side noticeably developed more than the other, particularly on the shoulders, on either side of the withers and on the croup? Is he happy to pick up a front foot when asked?

This is also a good time to ask if you can see the horse's passport and any other documentation and to check that the identification chart matches the horse you are looking at. At the same time, you can check his mouth and teeth to make sure he does not have marked conformation problems and, if you feel confident, to assess his age.

If you are happy with your first impression, ask the owner to bring the horse out and stand him up square on a hard, level surface. Stand back and review his conformation from both sides, front and back, first taking in an overall impression, then analyzing the parts which make up the whole. Make a mental note of anything that worries you and, if necessary, make a closer examination of any bumps or blemishes you noticed in the stable.

Pick up each foot in turn and notice if they seem well conformed. Unless he has been newly shod, look at the way his shoes have worn: is the wear fairly even, or is one side worn much more than the other?

Now ask to see the horse trotted up on a hard, level surface, making sure that the handler keeps the lead rope loose enough for him to carry his head naturally. Watch the horse walk away from you and then back towards you; ideally, you want to see a nice free walk where the horse overtracks – which means that the imprint of the hind feet overtrack those of the front feet.

Repeat this procedure in trot, which will give you the chance to see not only whether or not the horse is sound, but if he moves straight. Notice if he dishes on one or both front feet, or if he brushes either in front or behind.

By now you should have formed an impression of the horse. It's rare to fall in love at first sight, though it can happen occasionally, and you still need to make sure you carry on and make a methodical assessment. Usually, you will either decide that the horse could be a possibility and you would like to see more, or there may be something that puts you off completely.

Under saddle

If you are sure that the horse is not for you, it is better to say so at this stage rather than to take the visit further. As long as you do it tactfully – perhaps explain that whilst he is a very nice horse, he isn't quite the type you have in mind – a seller should not be offended. If you are hopeful, or haven't discounted the horse, ask to see him tacked up and ridden.

Never ride a horse that cannot first be shown off under saddle by someone else. For a start, you want to make sure that the reason no rider is available is not that the owner and everyone else on the yard knows it is likely to buck or take off; second, seeing a horse ridden gives you an idea of how it performs when ridden by someone who knows him.

Watch the horse tacked up and notice his attitude. Does he stay calm and relaxed, or does he put his head in the air and try to move away when the owner attempts to put on the bridle? Evasive behaviour such as this may mean that he associates being tacked up with discomfort, perhaps because he has been handled inconsiderately, or that his teeth need attention. Similarly, a horse who lays back his ears or tries to bite when the girth is being tightened may have been made uncomfortable by someone who has pinched him by girthing up roughly. Careful and patient handling will usually overcome such problems, but novice owners should really be looking for a horse with a straightforward attitude.

Look at the tack being used. A horse suitable for a first-time owner should go nicely in a simple bit and bridle, preferably a snaffle. It doesn't matter if the bridle is fitted with a noseband designed to prevent the horse opening his mouth too wide, but warning bells should ring if this is fastened so tightly he cannot open his mouth at all. It may be that the seller does not know how to adjust it correctly, but it may also be that the horse can be strong in some circumstances. If you suspect this may be the case, ask.

The seller should be happy to ride the horse for you at all paces and, if appropriate, to jump him. It makes things much easier if, to start with, this can be done in an outdoor arena with good footing, as you can observe the way the horse moves and how he reacts to his rider's instructions from a safe vantage point. However, not everyone has such a facility and you may have to watch the horse being ridden in an open field.

The horse should stand still whilst the rider gets on, either from the ground or from a mounting block. It is always best to use a mounting block, because it puts less strain on the horse and the saddle, but if the rider follows this method, check that the horse can be mounted from the ground if this is unavoidable. The seller may offer to demonstrate but, if not, this is something you can ask to see if you return with your advisor.

Allow the rider to warm up the horse before asking him to work properly, then see him worked in walk, trot and canter on both reins. You hope to get the impression of a horse who is happy in his work and obedient, who goes forward from one pace to the next willingly and is neither too lazy nor too onward bound.

If you intend to jump your horse, ask to see him ridden over some small jumps. They don't need to be big, because you are interested in the way the horse jumps, not the maximum height he can manage. Again, allow the rider to warm up, if necessary starting with poles on the ground, then jumping a small cross pole before moving on to small upright and spread fences. Ask to see the horse jumped towards the school entrance or field gate and away from it to see if he shows any reluctance when going away. If the horse is kept at a large yard with lots of facilities, you may be offered the chance to see him jump cross-country fences such as ditches and water.

By now you will know whether you would like to ride the horse, or if you feel he is not suitable for you. It is understandable to feel slightly nervous of riding a strange horse in front of its owner, but if you feel that he would be too much for your current level of ability, it is much better – and much fairer to everyone concerned, including the horse – to say so than to take risks. If you made it clear in your initial inquiries that you were looking for a first horse and did not exaggerate your riding experience, this hopefully will not arise. However, if it does, a sensible seller would rather you were honest than got into a situation where both you and the horse were unhappy or worried.

Hopefully, you will have liked what you have seen and be eager to try the horse. Try not to feel self-conscious about riding him in front of someone else and don't rush things; give yourself time to check the girth and make sure your stirrups are the correct length, then spend a few minutes in walk getting the feel of him. If he is taller, smaller, wider or narrower than the horse or horses you are used to, it will take a little while to get used to the different feel and to the different length of stride.

Find out if he answers your requests to start, stop and turn in both directions politely in response to clear but gentle aids, and when you feel ready and confident, ask him to go into trot. The horse will be getting used to you at the same time as you are getting used to him, particularly if he is used to always being ridden by the same person. Whilst, in theory, all riders use the same system of communication, there will be slight variations between one rider and another because of differences in posture, stability, length of leg and so on. You may need to experiment a little, perhaps by giving lighter or clearer leg aids, until you feel you are communicating on the same wavelength.

Again, ask for changes in direction and for transitions between trot, walk and halt. If the horse is well schooled, the owner may have demonstrated this to you by performing some lateral work, where the horse moves forward and sideways at the same time, or other schooling movements. Don't feel you have to do everything the seller did unless your riding is already at that level.

If you are still happy and confident, ask the horse to canter. If you give the right aids, he should strike off on the correct leg; if not, stay calm, come back to trot and ask again. The canter should feel controlled and well balanced, not that you are going faster than you want to. Incorporate transitions between the paces again and change direction; if you started cantering on the left rein, go forward to trot, change the rein and ask for canter on the right rein.

Anyone who has learned to ride at a good school and has reached the stage where they are competent enough to think of owning a horse will understand the terminology used here. If you are looking ahead but have not yet reached that level, it may sound as if learning to ride means learning a foreign language!

Riding on the right rein means the horse is going in a clockwise direction, either round the school or on a circle. Riding on the left rein means he is going anti-clockwise.

When a horse canters, he takes a longer stride with one foreleg than with the other; the leg which takes the longer stride is called the leading leg. For him to stay balanced, he should go into canter and lead with the right leg on the right rein and the left leg on the left rein. This is referred to as getting the correct strike off or the correct lead.

Does the horse work equally nicely on both reins, or does he feel stiffer on one than the other? Unless he is very well schooled and the rider is well balanced, a horse will often feel easier to ride on one rein than the other. It is something that can be improved, but a first horse's education should be well enough established for the difference not to be too marked.

Provided you feel happy and still feel that the horse is a possibility, you might want to try him over a few small jumps. Don't feel under pressure to jump the same fences as those he was ridden over during the seller's demonstration; at this stage, a small cross pole may be enough to give you the feel of how the horse jumps and whether he stays calm and willing on the approach, take-off and landing.

Even if you are convinced that this is the horse of your dreams, it is unwise to commit yourself on one visit, particularly if you have made the assessment alone. You also need to see how the horse behaves when he is hacked out. Many horses are more forward in open spaces, which is understandable, but you need to make sure that he is well behaved in traffic and is not nappy – another wonderful phrase which means that a horse is reluctant to leave others or to go away from familiar surroundings.

The seller may try to persuade you to agree to buy the horse on the strength of what you have seen and may tell you that other potential buyers are coming to see him. No matter how tempted you are, don't commit yourself at this stage. Instead, tell the seller that you are very interested and arrange to return as quickly as possible – ideally the next day and certainly within the same week – with your advisor.

On the road

It is vital that a horse is sensible on the roads, whether you intend hacking out to be your main activity or just part of your horse's regime. Even if you are lucky enough to have miles of off-road riding, you are still likely to meet farm and off-road vehicles plus hazards such as trail bikers and cyclists.

When you find a horse you think would suit you, you can either ask to hack out on him after completing the assessments detailed earlier, or wait until you return for a second visit with your advisor. Your return visit will probably follow the same format as your first, as your advisor will want to judge the horse's conformation, movement and temperament and see how he behaves in the stable and when ridden. If all goes well, ask the seller to take you for a short hack, if possible one where you can meet a variety of traffic and – if the location is suitable – have a short canter on a bridlepath or along the edge of a field.

It is unlikely that the seller will allow you to take the horse out on your own and, unless you know the owner, the horse and the area, it is not fair to expect this. After all, would you let a stranger take out your horse without supervision? Instead, you should be escorted by a competent rider who appreciates that you are trying the horse for suitability.

During the ride, make sure that you ride behind the other horse, alongside it and in front of it for short periods. Start off behind and stay there, if possible, until you have seen the horse's reaction to vehicles passing from both directions. Is the horse happy to go where he is asked, or does he seem more nervous when in front – or does he jog or pull when asked to go behind? If he tries to spin round or is reluctant to leave the yard, he has a schooling or behavioural problem that needs sorting out by an expert and is not a good prospect for a first-time owner.

Ideally, you will be able to canter the horse behind and in front of the other, but if this is not possible because of the lack of suitable ground, ask to go behind. Some horses become stronger in company and whilst this is often not a problem, you need to be confident you can stay in control.

If you don't know the area, ask your escort to tell you when you are about halfway round. You may be able to guess, as some horses suddenly develop an extra spring in their step when they hit the home straight! Again, this is not a problem, unless the horse starts jogging and misbehaving.

Make sure you ride in front for a while, then behind, just as you did on the outward journey. When you get back to the yard, you may want to tell your escort that you would like to ride past the entrance for a short way, with your horse in front. Although the horse may be surprised and show a little hesitation when you ask him to go past, he should obey your aids and not show marked resistance. If he does as you ask, ride a few yards farther on, then turn back.

By now you should know whether or not this is the horse you want to buy. If you have brought someone with you to help you make the decision, you need to discuss both your findings in private, either on the way home or there and then. Hopefully your advisor will support your decision to buy the horse; if so, you can move towards concluding the purchase.

Some people may suggest that you ask to have the horse on trial for a short while to see if you will get on with him. You could always ask, but don't be surprised if the owner refuses. Letting a stranger look after and ride a horse without supervision is a big risk and one that many sellers will not be prepared to take.

Agreeing a purchase

Once you have made your decision, you need to tell the owner that you would like to buy the horse, subject to a satisfactory pre-purchase veterinary examination. This is not carried out to find faults in the horse, but for a vet to give you an opinion on whether he is sound and suitable for the job you want him to do. Some people buy horses without having them vetted, but in most cases this is not recommended: if you get him home and find out that he has a problem, you can let yourself in for heartbreak and expense. No matter how honest or knowledgeable the seller, there are some problems that only a vet could identify, such as a heart defect.

This is also the time when you need to finalize negotiations on the price, if any. Haggling over the price is traditional, but not universal. Some sellers will always ask a little more than they hope to get – for instance, you may be able to buy a horse advertised at £3,250 for £3,000 – but others will tell you that they are not open to offers. There may still be some leeway; for instance, the seller may agree to deliver the horse for you as part of the agreement, or include some tack or rugs. Your advisor may help you, but the final decision and responsibility is yours.

When you agree to buy a horse, you make a contract with the seller. To protect both of you, it is best to pay a deposit; some sellers will accept a small amount whilst others will ask for ten per cent of the agreed purchase price. You should be given a receipt which identifies the buyer, seller and horse and states that a deposit has been accepted against an agreed purchase price, subject to a satisfactory pre-purchase veterinary examination, and that the deposit will be returned if the vetting is not satisfactory.

Dealers often have receipts that state their terms and conditions, so make sure these are acceptable and that you understand them;

for instance, they may say that if the horse proves unsuitable within ten days of purchase they will exchange it for one of similar value, but this does not necessarily mean that you will be offered a refund. A private seller is unlikely to offer any sort of refund, though someone who cares about a horse may buy him back rather than risk him being sold on to an unknown home.

13 vetting and insurance

In this chapter you will learn:

- **why vetting is important**
- **what the vetting procedure entails**
- **why you should insure your horse**
- **how to avoid insurance pitfalls.**

You need to arrange a pre-purchase veterinary examination as soon as possible after you have agreed to buy a horse. This is not a warranty, but rather the vet's professional opinion of its soundness and suitability for a particular job. Some sellers may show you the certificate issued when they had the horse vetted, but though this may be interesting, you can't assume that nothing has changed since then.

All vets are qualified to treat all animals, but most specialize in particular areas. It is important to make sure that the vet you consult is a specialist equine vet, who will be able to explain his or her findings to you and – as there is no such thing as a perfect horse – whether or not they would be likely to affect the horse for your use.

If you are buying a horse locally, a knowledgeable person such as your instructor should be able to advise you on which vet you could use. Things may not be so easy if you have found a horse in another part of the country; an equine practice local to you may be able to suggest someone, or you could ask the seller for suggestions and check before appointing anyone that the vet concerned is an equine specialist.

At one time it was thought unsuitable for a seller's own vet to examine the horse on behalf of someone else. However, if the vet is willing to do so and the seller is prepared for the horse's veterinary history to be revealed, this can work well. If the seller doesn't want you to know the horse's history, you have to ask yourself: why not?

You may also find that a seller does not want a particular vet to examine the horse, perhaps because they have had a disagreement in the past. Remember that it is up to you who you appoint, not the seller.

When you arrange the vetting, make sure the vet knows exactly what you want to do with the horse and at what level; there is a big difference between a riding-club novice one-day event and eventing at affiliated level, and a horse that is still suitable for one may not be as acceptable for the other. You should also mention any concerns that you have, however small – for instance, whether a blemish may be significant.

Try and be there when the vetting is carried out, if possible, but don't expect the vet to talk you through it stage by stage. Some will, but others prefer to concentrate on the task and discuss their findings with their client at the end. The advantage of being present is that the vet can actually point things out to you.

The only really satisfactory form of vetting for a riding horse is that which, in the UK, follows a recognized five stage procedure. The five stages are a preliminary examination; trotting up; strenuous exercise; a period of rest followed by a second trot up; and foot examination. Although vets have different ways of doing things, they will all work within this framework.

If you are buying an elderly pony as a safe mount for a child and he will not be expected to go out of trot, the vetting procedure may be shortened. It is still well worth asking the vet to examine the pony, as problems that may not be obvious, such as joint stiffness, eye or heart problems, can be picked up. Obviously it's disappointing to find out that such things exist, but it's better to find out than to put child or pony at risk.

Preliminary examination

Usually, this will be carried out in a stable. The vet forms an overall impression and will examine the horse from head to tail. He – supposing we have a male vet – will look at the horse's teeth to decide an approximate age and compare the horse he is looking at with the details in his passport to make sure they match; identity information includes not only colour, markings and any scars or blemishes but also whorls, patterns of hair on the head and body which are unique to every horse.

He will also examine the horse's eyes with an opthalmoscope to make sure there are no sight defects and use a stethoscope to listen to the heart and lungs at rest. These readings will later be compared to ones taken after the horse has been exercised.

If any part of the examination shows up a problem that means the horse is unsuitable, the vet will explain this to you and end the vetting there.

Foot emaination

At some stage of the vetting procedure, the vet will use a pair of hoof pincers to test for any sensitive areas in each foot in turn.

Trotting up

The next stage allows the vet to assess the horse's soundness and movement and he will ask the seller to walk and trot the horse away from him on a hard, level area. In most cases, he will also carry out flexion tests, where the limbs are held in a flexed position for between 45 seconds and a minute and the horse then trotted up immediately the limb is released. Whilst the vet may ignore a 'lame' stride or two as caused by the limb being held immobile rather than by an underlying problem, continuing lameness will be regarded as significant. This part of the examination underlines the importance of the vet's experience and ability to interpret his findings.

Strenuous exercise

The vet will now want the horse to be exerted so that he can listen to his heart and lungs and compare them with the readings he took during the initial examination at rest. The horse will be cantered for a suitable period – the amount of exertion will vary according to his current fitness – and the vet will also listen to the horse's breathing to make sure there is no untoward noise that may signify a respiratory problem.

Period of rest and second trot-up

By now the horse will have worked hard and he will be put back in a stable to rest for about half an hour. The vet will then ask the seller to walk and trot him up in hand exactly as before, to see if his findings are the same or if exertion has thrown up a problem.

He will also want to see the horse turned in his own length in either direction to make sure he can cross one hindleg over the other, and to back up. If the horse is reluctant or unable to do this, it may mean he has a problem.

The vet will listen to the heart and lungs again to check that they have returned to the normal base rates established on the preliminary examination.

Extra procedures

Optional extra procedures are available as part of the vetting procedure and, whilst some may apply mainly to expensive competition animals, there is one – taking a blood sample – which is recommended in all cases. The reason for doing this is that if a horse suddenly goes lame a few days after purchase, or shows a change in temperament over and above anxiety at being in a new home, the sample can be analyzed to rule out the possibility that the horse was given medication or other substances prior to the examination.

The sample will be taken by the vet and stored for a few weeks after the examination. It protects not just the buyer, but also the seller, and many dealers will insist that one is taken.

If the horse's breathing is noisier or more laboured than might be expected for its state of fitness, the vet may suggest carrying out an endoscopy. Colloquially known as 'scoping', this harmless procedure enables him to look at the upper part of the horse's respiratory system and to detect any abnormalities. X-rays may also be taken of the joints or feet, but this is not routine and X-rays require specialist interpretation.

Insurance

For most owners, some form of insurance is essential and this should be arranged as soon as the purchase is completed. If the horse has an accident or illness before you move him to his new home – even if it happens five minutes after you have paid for him – the cost and the consequences are down to you. Ideally, you should do some research as soon as you are ready to look at horses for sale and be ready to take out a policy as soon as you are officially an owner.

Some owners prefer not to take out insurance, but premiums are cheap compared with the cost of paying veterinary fees if your horse needs colic surgery – £3,000-plus – or having the money to buy another horse if you are unfortunate enough to lose him through accident or illness. Remember that veterinary costs will be the same whether the patient is a £500 pony or a £50,000 top-class competition horse.

Recent court judgements mean it is also important to take out third-party liability insurance. In the UK, this is included as a membership benefit of the British Horse Society but will also be available through your horse insurance provider.

There are many good specialist equestrian insurance companies and brokers; unfortunately, you only find out how good they are when something goes wrong. Although vets are not allowed to recommend specific providers, they do have a good idea of which ones are easy to deal with and pay promptly. It is also worth contacting some of the specialists who advertise regularly in the equestrian press and explain that you are about to buy a horse and need advice on insurance cover.

Some policies offer blanket cover for a number of areas whilst others allow you to pick the ones which apply to you. Compare costs, as whilst there are many good value general policies, you may find yourself paying for cover you do not need – for instance, if you keep your tack at home you may already be covered on your household policy for theft and damage to saddlery and tack.

All standard policies cover mortality, giving financial compensation if your horse dies or has to be put down on the instructions of a vet. Other areas include third-party liability and veterinary fees, as mentioned above; saddlery and tack and loss of use. As is common with most forms of insurance cover, you will be expected to pay an excess (the first part of any claim; amounts will vary).

Whilst third-party liability and veterinary fees cover can be regarded as essential, loss of use is very expensive. It means that if the horse becomes permanently unable to do one or more jobs for which he is insured, you will receive compensation – but it is an area fraught with complications.

Danger zones

Horse insurance is a complex area and there are potential pitfalls to catch out the unwary. When taking out a policy, bear the following points in mind.

- Compensation after the death of a horse or an agreed loss of use claim will be based on the sum insured or market value, whichever is the less. This is important: there is no point in trying to insure a horse for more than he is worth and a good insurer should not allow you to do this. It should generally be accepted that the price you paid is the market value.
- You will only be able to insure a horse for his actual, not potential, value. This applies particularly to younger horses: you may be convinced that you have bought a potential superstar, but you can only insure him on the basis of what he is now, not what he might be in future.
- The insurers will require you to fill in a proposal form giving details of the horse. This will include a declaration of soundness including statements that, to the best of your knowledge, the horse has never been operated on for certain conditions, such as colic. If you make a false declaration of any kind and are found out, the insurers will not pay out – and a convenient memory lapse could be construed as attempted fraud.

  If you have had the horse vetted, many insurers will want to see a copy of the certificate detailing the vet's findings.

You may find that if he has found evidence of an old injury – or even, in some cases, something such as a curb or splint – the insurers will want to impose an exclusion. This means that they will not be prepared to pay for any problem relating to it, or even to the whole of that limb.

However, a vet's certificate can also work in your favour. As insurers will not pay for pre-existing conditions – ones that could have been there when you bought the horse – it will often be accepted as evidence that a problem is a new one.

- Horses should be insured to take into account all the activities you intend to take part in. These are divided into categories according to their risk value. If you decide to take part in an activity you are not covered for and your horse has an accident, you will not be able to claim on your policy.

If you need to keep the cost of insurance premiums to the minimum, there are two ways of doing it – but you have to make sure that you are not making a false economy. The first strategy is to insure your horse for less than his market value, as the greater the value, the higher the premium. By doing this you have to accept that if the worst happens, you will not receive enough compensation to buy a horse of similar value.

The other strategy is to negotiate a higher excess on veterinary fees cover, as this accounts for the most expensive part of the protection offered. For instance, if the standard excess is £100 and you offer to pay up to £250 or even more, your insurers may be able to reduce your premium considerably. Some owners prefer to do this, accepting that they will be responsible for relatively small bills but still have cover for more expensive incidents such as surgery.

When taking out an insurance policy, check if it comes into force immediately or if there is a waiting period. Some insurers specify that cover will only apply after the first 10–14 days of ownership, or that it will only apply to accidental injury. As horses are more liable to suffer accidents or stress-related problems during the immediate change from one home to another, it is a risk that needs to be considered.

When you get your policy document, read it immediately and thoroughly. Most companies have made their policies more user friendly, but they are still legal documents. If there is anything you don't understand, ask for an explanation and if there is anything you find unacceptable, cancel it during the period specified in the policy and find alternative cover.

14 taking your horse home

In this chapter you will learn:

- **how to transport a horse safely**
- **how to settle him in to a new home**
- **how to start a successful partnership.**

Getting your new horse home is exciting and also nerve racking, but there are lots of ways to make sure you start your new partnership in the best way. Before you make arrangements to transport him, get the seller to write down details of what and how much he has been fed, the routine he is used to and anything they can think of that will enable you to make a change in owner and environment as free from stress as possible.

The seller may be prepared to deliver him, though a dealer may charge extra. If this is not possible, perhaps because the distance is too great, you need to arrange for someone competent and experienced to collect him. Don't try and do this yourself with borrowed or hired vehicles unless you are already experienced and confident. Even though you may be able to legally tow a trailer or drive a non-HGV horsebox on your car driving licence, you should not assume this, as there are special regulations which apply. Also, you should not attempt this until you have learned and practised the necessary skills.

Some people assume that because they can tow a caravan or have driven a small lorry they are automatically capable of transporting a horse, but this is not the case. Transporting a live load that can and will shift its weight and, if the driver is not skilful enough, lose its balance is not fair on the horse, and unless you have been taught by an expert you could be more liable to have an accident.

Be careful, too, if a friend offers to transport the horse for you. If they do it without any money changing hands, there should not be a problem, but if you pay them a fee, they are using their vehicle for hire and reward. This means that they will probably be contravening their insurance and will be breaking the law.

Many new owners will find that their best option is to use a professional licensed transporter. Although this may seem expensive, you know that the vehicle will be safe and the driver will be experienced both on the road and in dealing with horses.

Travelling gear

Horses must be properly equipped for travelling to keep them safe and comfortable. Always use a leather headcollar rather than a nylon one, as leather will break in an emergency whereas nylon often won't and the horse is more likely to be injured. The lead rope attached to it, which will be used to tie up the horse in the vehicle, should be in good condition with a good quality clip at the end. It's a good idea to take a spare headcollar and lead rope just in case one breaks.

Some owners like to use a pollguard for extra protection, especially on a big horse whose head will be nearer to the roof. This is a shaped pad which attaches to the headcollar and protects the sensitive poll area. If it has earholes, make sure they are large enough to prevent your horse's ears being pinched.

Your horse will need to wear a rug suitable for the weather conditions and, depending on the time of year, whether or not he is clipped. All vehicles should be well ventilated and it is better to put on a light rug than to close air vents. In cold weather, a lightweight thermal rug is often a good option, whilst in hot weather, a cotton summer sheet or breathable cooler rug should be suitable. A horse can't move out of the way of draughts in transit, so should always wear at least a lightweight rug to prevent his muscles getting chilled.

Horses' legs are vulnerable to injury when travelling, as they may shift their balance and knock one against the other, so need to be protected either with specially designed travelling boots or bandages over padding. Boots are quick and easy to put on and there are designs which cover and protect the legs from the knees and hocks down to the coronet band at the top of the hoof. However, unless they fit perfectly, boots may be more likely to slip than correctly put on bandages; ask the seller what the horse is used to wearing.

Travelling bandages are wider than tail bandages and must usually be used over pads, though there are some which incorporate padding within their length. If in doubt, put them on over bandage pads – which, like all travelling equipment, will be available from your local saddler.

Many people wrap the pad round the leg first and hold it in place whilst they apply the bandage, but there is an easier way. As with tail bandages, make sure that the leg bandages are wound correctly so that when you get to the end, the fastenings are on the outside, not the inside. Then unwrap a short length of bandage, place it about 3 cms below the top edge of the pad and wrap both round the horse's leg, continuing the bandage all the way down and back up again. This method makes it easier to hold the pad in place.

Your bandages should start just below the knee and hock joints and go down over the fetlocks before returning back up the legs, with the pads showing above and below. Keep the tension firm and even and secure the fastenings on the outside of the leg, not at the front or back, where they could press on bone or tendon. Some people like to use knee and hock boots as well, but these have to be fastened tightly at the top to stay in place and can cause more problems than they prevent.

A tail bandage, applied as explained in Chapter 06, will help protect the tail hairs from being rubbed if the horse leans back in the vehicle. A padded tailguard which fastens with tapes or Velcro/hook-and-eye fastenings gives extra protection.

Make sure you take an equine first aid kit with you; hopefully it won't be needed, but you should never travel a horse without one. Before you load up and set off, make sure that you have your new horse's passport and any other documentation and that you have a receipt for the full amount paid. Some horses travel better if they have a hay or haylage net to pick out, but make sure it is tied up securely.

If you're making a long journey and think the weather might change, take a selection of rugs, together with a water bucket and container so that you can stop halfway and offer him a drink. Fill the container at the horse's yard, as the taste of water varies from one part of the country to another. This also allows you to mix 'old' water with 'new' when you get him home, which will encourage him to drink. Some horses don't mind, but others are suspicious of water which tastes different and you need to make sure he does not become dehydrated.

At the yard

Ideally, arrange for your horse to arrive during daylight hours, which will make it easier for him to start getting used to his new surroundings. Unless he is going to live out all the time, it's usually best to take him straight to his stable, which should already have been prepared with a clean bed and hay or haylage waiting and a supply of clean, fresh water. This gives you chance to take off his travelling gear and check that he hasn't banged or scraped himself whilst travelling. Try and make sure that he can see at least one other horse, as many horses will become agitated if they find themselves on a new yard without equine company.

Inevitably, other owners on the yard will want to come and see your new horse, but allow him time to settle. Horses vary in their reactions; some are visibly anxious, whilst others seem relatively unconcerned. No matter how laid back he seems, don't take things for granted; although he might not show it, he will still be responding to new people and places and may react quickly to noises or movement.

If all seems well and there is enough time, turn him out in the field so he can have a roll and hopefully graze. The ideal way to introduce a new horse to an established group is to turn him out in a small paddock next to them with a quiet companion. All groups have a pecking order and your horse will have to find his place within it; if you can turn him out with a companion who is peaceable and fairly low down in the order, you should soon be able to put both of them in with the main group.

No matter how careful you are, there is always a certain amount of posturing when a new horse arrives. Don't worry if the other horses squeal at him, lay back their ears or seem to ostracize him and keep him on the outskirts of the group. Eventually, one of them will accept him and he will be admitted to the group.

If he has had most of the arrival day to get used to his surroundings, it's generally best to get on him for the first time the next day; some people might advise you to give him several days to settle down but, if he is used to being worked every day, this will mean that he may become less attentive or even a bit too full of himself.

Don't take risks. If possible, turn him out first and ride him later in the day; if this is not possible and you feel you want to make sure he is settled enough to ride safely, fit protective boots on him and lunge him for 15 minutes in an enclosed area first. You might feel more confident if you can arrange for your instructor to be there the first time you ride, so that you have someone experienced to help you gauge his reactions and cope with any communication problems. If you can't do this, follow the same sort of routine as when you tried him, concentrating on getting to know him and establishing communication rather than asking for anything complicated.

You might find that 20 minutes in the school is enough and you can then either turn him out again or put him in his stable with plenty of hay. Don't feel as if you are embarking on a great challenge – you have lots of time and need to get to know each other. At the same time, you need to be relaxed and positive in your riding, because your horse will take confidence from you.

As soon as possible, find someone reliable, with a quiet horse, to hack out with you. Aim to go out for about an hour and try and pick a route that is usually as free from hazards as possible. Your horse will take confidence from the other horse and from you and, every time you ride, you will learn more about each other. Once you feel that your horse has settled, you can hack out alone, if necessary.

Safety on the roads

If you are used to riding out in groups from a riding school, you will hopefully be aware of the importance of road safety. It is just as vital when riding as an individual, so you might want to remind yourself of the following guidelines:

- High-visibility, reflective, fluorescent gear for horses and riders means you are more easily seen, from greater distances. This gives other road users time to see you and slow down.
- Ride on the correct side of the road – which, in the UK, is on the left – and if you are riding with a companion, don't lose concentration on your horse and your surroundings because you get too involved in conversation.
- Be as courteous to other road users as you hope they will be to you. Thank considerate drivers with a smile and a nod – you don't need to take your hands off the reins.
- Make sure you know the appropriate hand signals for turning left and right at junctions and roundabouts, and for requesting drivers to slow down or stop.
- Be alert for possible hazards and look behind you at regular intervals so you are aware of approaching traffic.
- Don't move out unless it is safe to do so. Look behind you, give the correct hand signal and, if it's safe, carry on. If it isn't safe, perhaps because a car is overtaking you, wait.
- Always carry a mobile phone, but keep it swiched off. Riders who have their reins in one hand because they are using their phone with the other are as dangerous as drivers and may be just as liable to prosecution. Make sure your phone contains your personal ICE (in case of emergency) number plus those for your vet and home/livery yard.

Going it alone

Many first-time owners find that one of the most bewildering things about becoming a horse owner is that when you are riding, you suddenly have to think for yourself. There is no one to tell you to change the rein or ride a circle at C or to dictate on a hack when you are going to trot or canter. So whilst riding should be a pleasure and a relaxation, it should never be mindless: before you get on your horse, decide what you are going to do and what you want to achieve.

Perhaps you want to achieve a quicker response to your aids, improve the smoothness of transitions within paces, or work on an aspect your instructor has suggested during a lesson. If you want to progress, regular lessons are just as important now you own a horse.

Don't be disheartened if things occasionally go wrong, which they will: you're developing a relationship with another living being, not a bicycle, and there are bound to be days when you have misunderstandings. The important thing is not to ignore small problems in the hope that they will go away, but to get advice in solving them before they turn into big ones.

You'll find that there is a whole new world out there now you own a horse of your own, whether you want to compete or ride purely for pleasure – though the two can and should go together. And as the months go by, you'll discover that one of the great things about owning a horse is that you never stop learning.

Useful websites

Association of British Riding Schools
www.abrs.org

British Association of Equine Dental Technicians
www.equine dentistry.org.uk

British Equine Veterinary Association
www.beva.org.uk

British Horse Society
www.bhs.org.uk

Farriers Registration Council
www.farrier-reg.gov.uk

International League for the Protection of Horses
www.ilph.org

Society of Master Saddlers
www.mastersaddlers.co.uk

Useful organizations

Association of British Riding Schools
Queen's Chambers, 38–40 Queen Street, Penzance, Cornwall, TR18 4BH (Tel: 01736 369440) www.abrs-info.org

Blue Cross Equine Centre
Shilton Road, Burford, Oxon, OX18 4PF (Tel: 01993 822651) www.bluecross.org.uk

British Association of Equine Dental Technicians
www.equinedentistry.org.uk – cannot answer telephone queries from owners but includes a list of qualified members.

British Equine Veterinary Association
www.beva.org.uk – cannot answer individual queries from owners but contains general advice on horse health.

British Horse Society
Stoneleigh Deer Park, Kenilworth, Warwickshire, CV8 2XZ (Tel: 08701 202244) www.bhs.org

Farriers Registration Council
Sefton House, Adam Court, Newark Road, Peterborough, PE1 5PP (Tel: 01733 319911) www.farrier-reg.gov.uk

International League for the Protection of Horses
Anne Colvin House, Snetteron, Norfolk, NR16 2LR (Tel: 0870 870 1927) www.ilph.org

Society of Master Saddlers
Green Lane Farm, Stonham, Stowmarket, Suffolk, P14 5DS (Tel: 01449 711642) www.mastersaddlers.co.uk

Suggested reading

Robert Oliver and Bob Langrish, *A Photographic Guide To Conformation*, (JA Allen).

Carolyn Henderson and Lynn Russell, *How to Buy The Right Horse*, (Swan Hill Press).

Karen Coumbe, *First Aid For Horses*, (J A Allen).

Carolyn Henderson, *Tack: How To Choose It And Use It*, (Swan Hill Press).

Ruth Bishop, *The Horse Nutrition Bible*, (David and Charles).

John Henderson, *The Glovebox Guide To Transporting Horses*, (J A Allen).

index

teach yourself

beekeeping

adrian and claire waring

- Would you like to produce your own honey?
- Do you need information about equipment?
- Would you like advice on seasonal needs?

Whether you are simply curious, or are an amateur beekeeper already, **Beekeeping** will teach you everything you need to know. From bee biology to the best hoves for town or country bees, covering courses, up-keep, and even dealing with the neighbours, this friendly introduction has information on taking your hobby further, and even includes some great honey recipes.

Adrian and Claire Waring have over 65 years beekeeping experience between them. Adrian is a former County Bee Instructor and Examiner and Claire is Editor of *Bee Craft*. Both are former General Secretaries of the British Beekeepers' Association.

teach yourself

birdwatching

wildfowl and wetlands trust

- Are you a beginner or novice bird-watcher?
- Do you want hints on where to go and what to look for?
- Do you want guidelines on different species and varieties?

If you have just started birdwatching or think you might like to know more about the hobby, this book will tell you everything you need to know. From equipment needs to the different species, with guidelines for what to do and what not to do, and even how to make birds come to you and your garden, **Birdwatching** features a bird for every month, and gives ample resources, including checklists and top tips.

Founded in 1946 by the artist and naturalist Sir Peter Scott, the **Wildfowl and Wetlands Trust** is the largest international wetland conservation charity in the UK, with nine visitor centres around the country.

teach yourself

dog training

association of pet dog trainers

- Do you want a comprehensive guide to training your dog?
- Would you like your dog to be socially well-behaved?
- Do you need advice on all aspects of being a dog owner?

If you want your dog to be well-behaved then **Dog Training** is for you. Essential reading for all dog owners or those thinking of buying a dog for the first time, this book covers every aspect of kind, fair and effective dog training as well as authoritative advice on looking after your pet. Using positive, reward and motivational techniques, including clicker training, you will be able to train your dog to be obedient, sociable and, most importantly, to be a part of your family.

Association of Pet Dog Trainers offers pet dog owners a guarantee of quality when looking for dog training advice. The APDT abide by kind and fair principles of training and have written this book accordingly. For more information, visit www.apdt.co.uk

teach yourself

keeping poultry

victoria roberts

- Do you want to know which breed lays best?
- Would you like advice on housing and equipment?
- Are you considering keeping ducks and geese?

Whether you want to start from scratch with a few hens, or branch into ducks, geese and other birds, **Keeping Poultry** is for you. It tells you which breed of bird lays best and gives useful guidance on housing, equipment and the necessities of day-to-day care. Covering all types of poultry, this guide offers advice on everything from exhibiting birds to meat production, with a full 'trouble-shooting' section and even tips for breeding your birds.

Victoria Roberts, BVSc MRCVS, is the author or editor of five books on keeping poultry, the Honorary Veterinary Surgeon for the Poultry Club, and the Editor of the Poultry Club newsletter.